CLOCKS
CONSTRUCTION
MAINTENANCE & REPAIR

CLOCKS

CONSTRUCTION MAINTENANCE & REPAIR

BY FRANK W. COGGINS

TAB BOOKS Inc.
BLUE RIDGE SUMMIT, PA. 17214

FIRST EDITION

FIRST PRINTING

Printed in the United States of America

Library of Congress Cataloging in Publication Data

Coggins, Frank W., 1919-
Clocks—construction, maintenance, and repair.

Includes index.
1. Clock and watch making. 2. Clocks and watches—Repairing and adjusting. I. Title.
TS545.C63 1984 681.1'13 83-24161
ISBN 0-8306-0269-0
ISBN 0-8306-0169-4 (pbk.)

Front cover photograph courtesy of Brookstone Company, 1091 Vose Farm Road, Peterborough, NH 03458.

Contents

Acknowledgments

MORE THAN THREE YEARS WENT INTO MAKING this book. My heartfelt thanks go to each and every person who said, "If this is what you want to do, how can I help?"

There is no way, whatsoever, that this work could have been completed without the constant, dedicated assistance of my wife, Nancy. Even more remarkable is the fact that her unflagging efforts with research, her cheerful forays over three states in the replenishment of supplies and materials, and her never-empty coffee pot have helped me through the completion of three other books as well!

It would be virtually impossible to compile a list of names of all the people who provided a hand in this endeavor. Notable among them are Nancy Miller, who did most of the drawings; Marc and Marion Behr; Nora Edwards, James W. Gray, William C. Temme, Caroline Herman, Dorothy Ashton of Creamer Dickson Basford, Inc.; Loretta Holz, Joe Oldham, Burton Unger, Evalyn Carter, Marjorie Coggins (who even made available valuable photographic equipment), Joe Zabinski, and Mr. Adrian B. Lopez.

Special thanks to the many manufacturers who furnished tools, equipment, and information. They include the Seth Thomas Clock Company, Dremel Manufacturing Co., Sears, Roebuck and Co., Bordens, Inc., Krylon, Stanley Tool Co., Burgess Tool Co., Lorraine Gary and Richard Hillman, Erika Gordon, The Old Bridge Library, Old Bridge, NJ, Susan Krieger, The 3M Company, Don Ellenberger, Patricia Hayes, and Lillian Smith.

I would like to express gratitude to the many people who commissioned timepieces or the repair thereof and then granted permission for the pieces to be discussed or photographed for use within these pages. They include Don Chaiken of *Mechanix Illustrated*, Marc Behr, Susan Sawicki, Gerri Miller, Matilda Jimenex, and Nancy Miller—to name a few.

Unless otherwise specified, the photography is by Nancy Coggins and Frank Coggins.

Introduction

THIS BOOK IS INTENDED FOR EVERY SINGLE person who has had an alarm clock refuse to go off in the morning and, rather than toss it out the window, decided to repair it. That a timepiece might suddenly self-destruct, with springs and wheels (they are *not* called gears) flying all over the place, might be a surprise, but then how few of us realize just how much energy is stored in a mainspring!

This book also is intended for craftspeople, no matter what their media, who are looking for items to make that have a quality of longevity about them and, as gifts, will delight the recipient. For people who shy away from long, seemingly involved projects that are basically the construction of beautiful pieces of furniture to which is added a completely assembled and ready-to-use clockworks, there are included here projects that will offer a true feeling of accomplishment.

There are more than 50,000 people in the United States who are so involved in horology that many of them have formed an organization, The National Association of Watch and Collectors, Inc., to share interests and swap ideas and timepieces. To some of them, facets of what is offered herein will be nothing more than an abomination! I admit, for example, to having cannabalized some electric clocks earlier on which might well be considered collectibles today. I hope to steer you around this pitfall.

The state of the art of designing and manufacturing timepieces moves on at such a rapid pace that today "little old watchmakers" find themselves incapable of keeping up with repair technology. This sad state of affairs (for devoted horologists) is not quite as serious in the case of clocks. Key-wind clocks, for example, are still manufactured and sell well—particularly in the case of the more expensive models. Battery-powered quartz, liquid crystal display, and light-emitting diode mechanisms are fast superseding ac-powered clocks. It is estimated that in less than 10 years, modern clock technologies will have just about done away with anything that does anything other than hum almost inaudibly. Yet there remains a wealth of building, repair, and maintenance pleasure to be had in the existing key wind, ac, and typical battery-powered types for the enthusiastic tinkerer or the avid horologist.

When a clock fitted with chiming and striking mechanisms is examined cursorily, it seems extremely complicated. Yet it does not take years of study or investigation to discover how misleading this first glance can be. A look at the workings of an anniversary or 100-day clock (even with its fancy revolving pendulum) will belie this. The brass wheels with which these mechanisms are fitted are a joy to behold. A large portion of this book is devoted to exposing these wheels in modern versions of what was first called *skeleton clocks* in the 1850s.

Do not assume at this juncture that the comforting tick of a clock or the hands in front of the dial are doomed to as early and total a demise as I have suggested. LEDs do have drawbacks, and chief among them are how difficult they are to read at night. An article in *The New York Times,* early in 1982, pointed out that mechanical timepieces, particularly watches, will continue to be used in undeveloped nations for a long time to come because the batteries and technologies to repair the modern offerings might not be so readily available in the immediate future. At the other end of the spectrum, no one knows just what might happen to a liquid crystal display mechanism if it is exposed to the extreme pressures and temperatures encountered

Fig. I-1. The Colgate clock is the world's largest timepiece. It is situated in Jersey City, New Jersey.

Fig. I-2. A typical stanchion clock, many of which can still be seen in cities throughout the United States.

in outer space. In other words, astronauts still wear wind-up watches—made by Patek Phillipe or Rolex, perhaps, but still requiring the thumb and index finger to keep them running.

There are some clocks with which this book will not deal except, perhaps, in citing their contribution to the 400-year history of timekeeping. Steeple and tower clocks are beyond the scope of this book. The basic information supplied here, however, might enlighten you enough to make a repair on such a clock if you have the tools and the physical stamina to climb into a belfry. Winding such a timepiece is a task in itself. That is why so many steeple clocks have been converted to electric motivation.

Just imagine the layman approaching the Colgate clock (Fig. I-1). It is the world's largest timepiece; the minute hand is 27 feet long and the hour hand measures 19½ feet. The combined weight of the hands and movement is about 4 tons. It is said that the tip of the minute hand travels something like three-quarters of a mile a day. Fancy the hazards in messing with such a formidable work.

There is a story about an English clock repairer who lost his right hand when adjusting a clock not nearly as large as this. The minute hand came around when the rope holding a weight broke. No rehabilitation programs existed in those days, and that chap was relegated to a life of poverty.

Large grandfather clocks or tall-case clocks border on this category, as do those clocks on stanchions (Fig. I-2) so prevalent in the streets of older cities throughout the United States. Older tall-case clocks frequently require that you actually machine the wheels and parts required for repair. It might prove feasible in such a case to replace the whole mechanism if the piece has some nostalgic value or simply not touch it if it has become a valuable collector's item.

Should a hint of irreverence or a lack of veneration be detected in the approach to horology in this book, suffice it to say that the love here is not unlike that developed by some of the most famous manufacturers of clocks in the United States during the 1800s. Not a few of them started making timepieces because they had a brass rolling/stamping mill and sought to expand their product line! Others went into the business as investors, among them P.T. Barnum who affiliated himself with the Ansonia Clock Company in the early 1850s. When the factory burned to the ground in 1854, he formed a merger with the Jerome Manufacturing Company (deliberately underestimating his debts in so doing). By the time the Jerome Company discovered that his indebtedness was three times greater than he had stated, Chauncey Jerome and his son had been held liable for the debt. Barnum then had the unmitigated gall to transfer his assets and property and cry that he had been dealt a disastrous financial blow by the Jerome Company! One of the principals in this fiasco committed suicide and Chauncey Jerome, whose clocks are highly prized and sought after today, died almost penniless.

All of the excitement of horology is not in the collecting of rare clocks or in the construction of ornate cases in which to house timepieces so beautiful in themselves. Much of it lies in reactivating a piece that has lain silent for years. More of the excitement comes from embracing theories and principles that reach back hundreds of years and in being able to incorporate them into something live with a feel of "today" about it, yet with a potential of being used in excess of 50 years!

The mantel clock shown in Fig. I-3 was purchased for $50 in a little antique shop in Connecticut. The proprietor felt that the plates had to be

Fig. I-3. A mantel clock, made by Waterbury Clock Company, and sold through the Sears, Roebuck catalog for $6.30 in 1897 (courtesy of Nora Edwards).

rebushed to make it run. It was put in the trunk of our car for the trip back to our studio and shortly after starting the trek home it was heard to chime. It was running when taken from the trunk of the car. The mechanism was removed from the case, given a thorough cleaning, reinstalled in the case, and it has been running for more than a year. The wear on the pivot holes was minimal, and certainly not bad enough to justify rebushing. The clock was made by the Waterbury Clock Company, Waterbury, Connecticut, and sold in the 1897 issue of the Sears, Roebuck and Company catalog for $6.30. It would sell today, as it stands (or ticks), for somewhere between $200 and $500.

This, is what this book is all about! Start from scratch, if you will, to make a clock. There are helpful hints herein. Don't hesitate to do something about that cuckoo clock that everybody in the family loved before it was discovered that the weights could no longer be raised. Cuckoos can be fixed, and all their secrets are laid bare within these pages.

Don't forget the alarm clocks. All is not lost if you purchased one in a hotel gift shop to wake you up the next morning and then promptly overwound it. There's a fix for that here, too.

If you must have everything (all the components) in kit form, there is no stigma to this. Some of the kits included will prove most helpful in determining for the future just how the energy from a coiled spring winds up turning two, three, or more hands with such surprising accuracy.

For those of you who would have a tower clock—or any of the others made down through the years by makers who honestly felt they were contributing to the advance of civilization—a comprehensive bibliography has been provided. I am most indebted to chaps like Charles Terwilliger and Henry B. Fried, whose writings helped me along the rocky path to the satisfactions derived from tinkering with clocks, and I pass them on to you.

Chapter 1

Tools and the Workplace

SERIOUS INVOLVEMENT WITH CLOCKS PRE- cludes leaving work in progress on the dining room table. Provision must be made for a somewhat isolated work site. There must be adequate lighting and, most definitely, a place for the storage of tools and parts. Considering that you might spend hours squinting through a jeweller's loupe, the workbench should be at least 3 feet high with a stool to match. A table or desk 30 inches high need not be rejected out of hand, but it will soon be discovered that a clock-regulating stand or a special clock vise or clamp will be required to raise the workpiece several inches.

This bench should have two or three drawers on the right side for tools and items that are used constantly. The top should be smooth, uncluttered always, covered with Formica or a large piece of glass. All the primers on clock repair, without exception, recommend that a 2 inch edging to be placed around the outer edge of the bench on three sides to keep screws and other small parts from falling onto the floor. You should work quite close to the bench for just this reason. It is amazing how far a tiny nut or screw will roll and bounce if it should slip from tweezers to the floor.

A glance around the work room (Figs. 1-1A and 1-1B) in which this is being written reveals a single pedestal desk with a wooden top 60 inches long by 30 inches wide. To it is clamped on various occasions one or two different types of vise. Permanently mounted in a draftsman's two-bulb fluorescent light that is supplemented by a fixture hanging from the ceiling with 2-foot fluorescent bulbs. This might seem a bit like overkill, but the hanging fixture illuminates the cabinets, credenzas, and cases surrounding and behind me in an L-shaped configuration. This arrangement, in one corner of the room, leaves plenty of space elsewhere for the display of clocks, for book shelves, file cabinets, and even one or two comfortable chairs.

Some people work on a shelf permanently af-

Fig. 1-1A. A workroom with a "bench" on the right. Lighting is fluorescent with incandescent for writing.

Fig. 1-1B. A view of one wall in the work area.

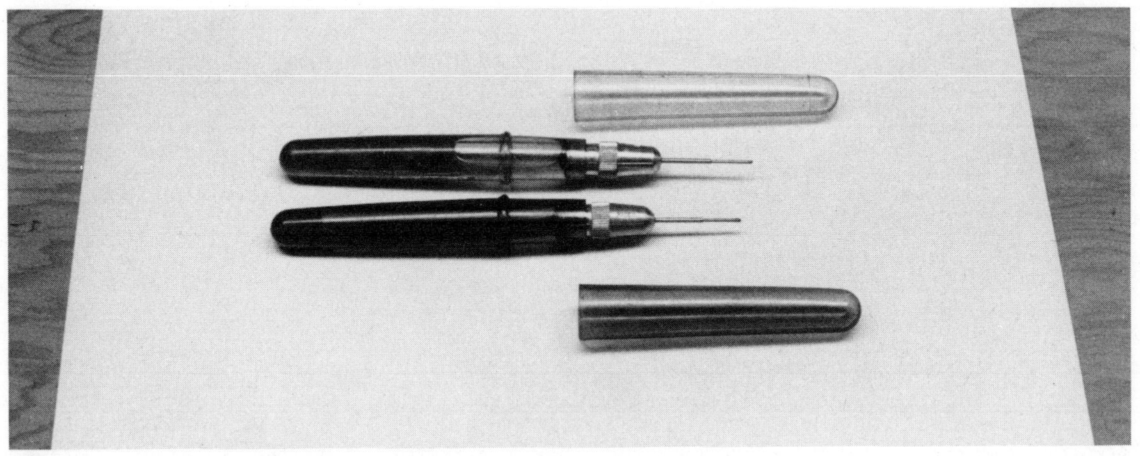

Fig. 1-2. A pair of clockmaker's oilers.

fixed to a wall; that is quite acceptable to those who do not object to facing a wall constantly.

I happen to be totally averse to someone coming upon me from behind, unexpectedly, when I am completely engrossed in some delicate work.

TOOLS

No work will go forward very effectively without a battery of tools. (Figs. 1-2 through 1-7). Once you are committed to clocks, it is a very short time before the hand tools, the screwdrivers, and the pliers already at hand become most inadequate and must be supplemented. A list of tools that will make life much easier follows:

☐ Several pairs of pliers: flat-nosed, round-nosed, parrot-nosed, and eventually, brass-faced flat-nosed.

☐ Screwdrivers (start collecting them!) in various blade widths and lengths. Include here a set of watch or jeweller's screwdrivers.

☐ Tweezers, fine and heavy and at least one with curved points.

☐ Hand pullers (to remove clock hands without touching the dial).

☐ Oilers (at least two).

☐ Cleaning brushes, hard, soft, some with brass bristles, some shaped like toothbrushes.

☐ Punches.

☐ Pin-vise.

☐ Regular vise with 2- to 4-inch jaws.

☐ Broaches.

☐ Files; needle files (available in sets) are very necessary. Swiss Pattern files in 5- and 6-inch

Fig. 1-3. An assortment of clock-winding keys.

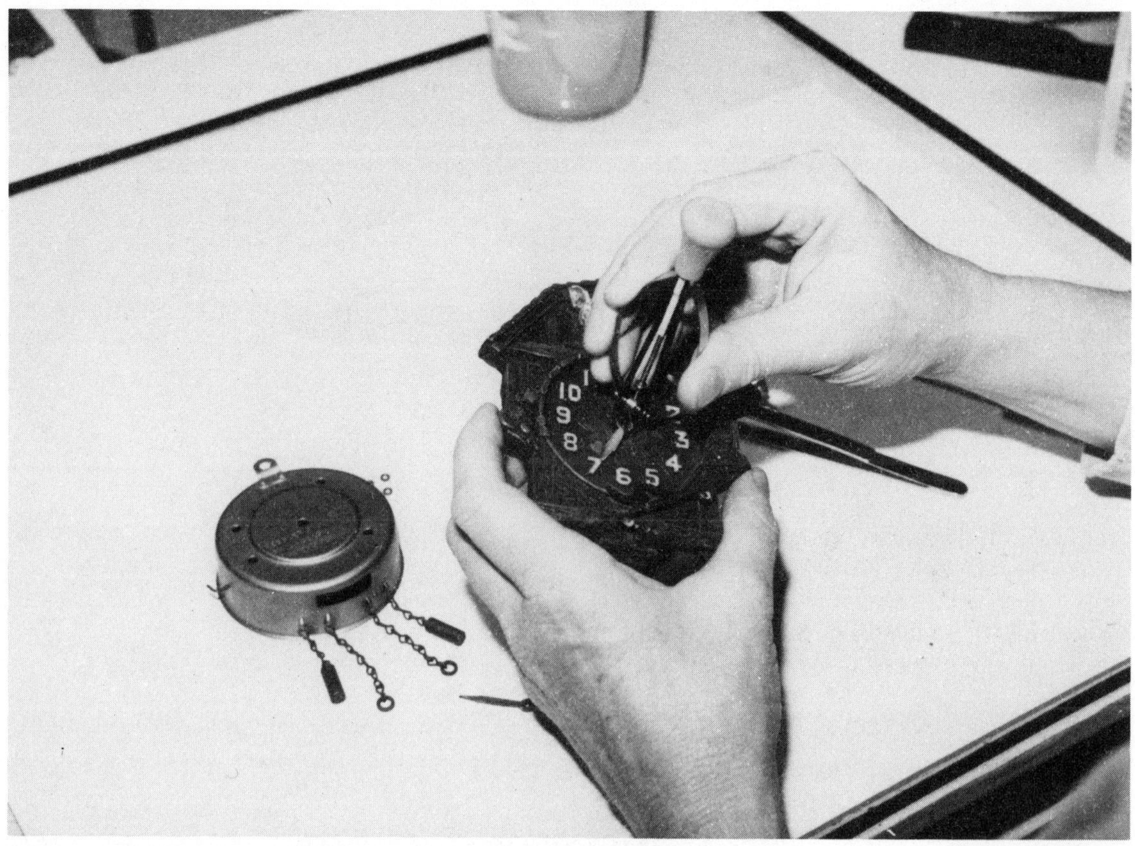

Fig. 1-4. Hand pullers for removing clock hands without touching the dial.

lengths should also be acquired. Select a half-round, triangular (3-Square) and a Round (Rat tail), all with handles if possible.

☐ Nippers, side and top cutting.

☐ Jeweller's loupes, 2½-inch and 3½-inch focus.

☐ Hammers, clockmaker's, round-face plus one brass-faced is available.

☐ Bench keys for letting down mainsprings.

☐ Clock keys, 5-prong

☐ Micrometer.

☐ Clock-cleaning solution.

☐ Dremel Moto-Tool or similar rotary device with accessory grinders, buffers, brushes, cut-off wheels, etc.

☐ Soldering iron.

☐ X-Acto knife with a variety of blades.

The longer one works with clocks the more frequent use each and every one of these tools will get. Months might pass without the micrometer being used, but as in the case of repairing the anniversary, described later, the job couldn't have been done without it.

There is little call these days for jeweller's lathes or turns. The need to cut wheels, pivots, or pinions is so rare that this service can be purchased. Ads for this type of trade work appear in *The Clock & Watch Review* and in the *Mart* of The National Association of Watch and Clock Collectors. There is a relatively inexpensive lathe, made by Dremel, that can be used for a small portion of this type of work. Clever use of Dremel's Moto-Tool will result in the satisfactory reduction of an arbor end or a pivot that will keep a clock running until it can be replaced. One could work for years without more than a wishful glance at a picture of a Boley or a Maximat.

THE MOVEMENT

It has been said that "fools rush in where angels fear to tread." Children are quite like this as is exemplified by the case of a youngster who took a large clock movement and made a working four-wheeled vehicle out of it. Whether or not he knew the difference between an escape wheel and a pinion is a moot question, but he did grow up to be one of the largest manufacturers of snowmobiles.

For the purposes of this book and all the projects that follow, some terminology must be injected which will be used from cover to cover. A few are listed here, more will be included in the Glossary, and still others will be defined as they appear in the text.

Balance. A vibrating wheel that, with a hairspring, regulates the motion of the going train in a small clock or watch.

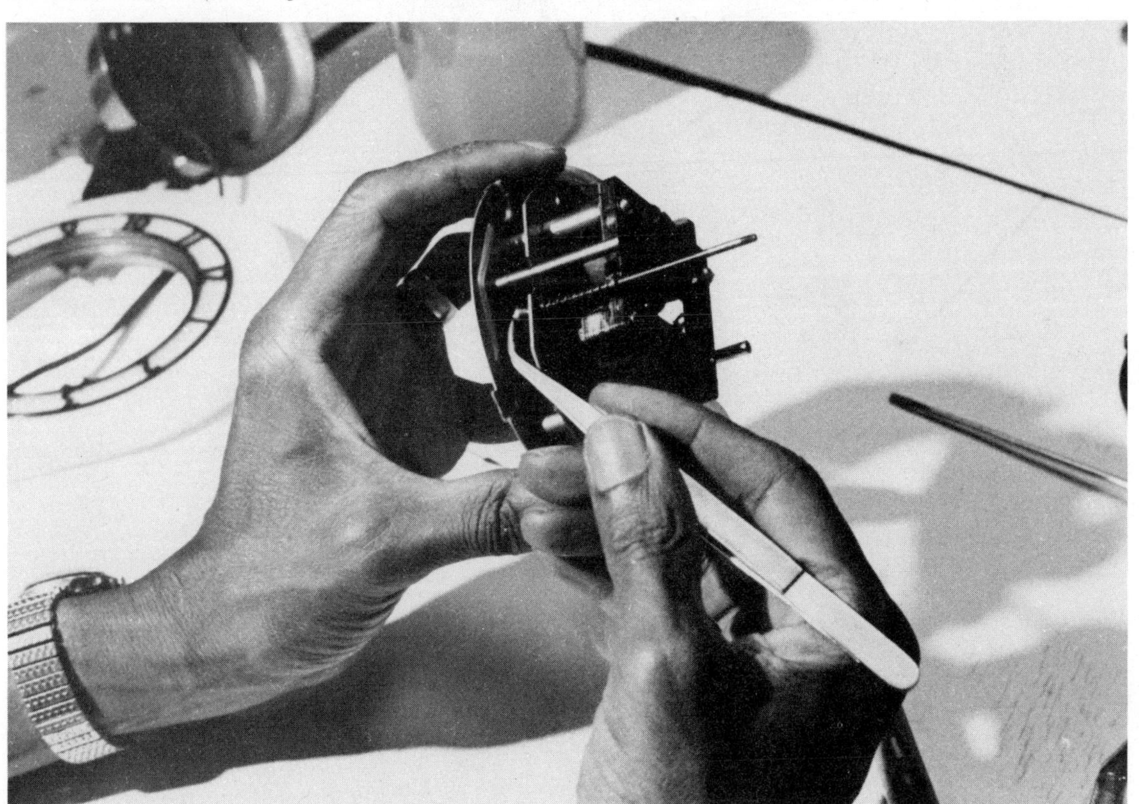

Fig. 1-5. Tweezers in use. The operator should have several different types on hand.

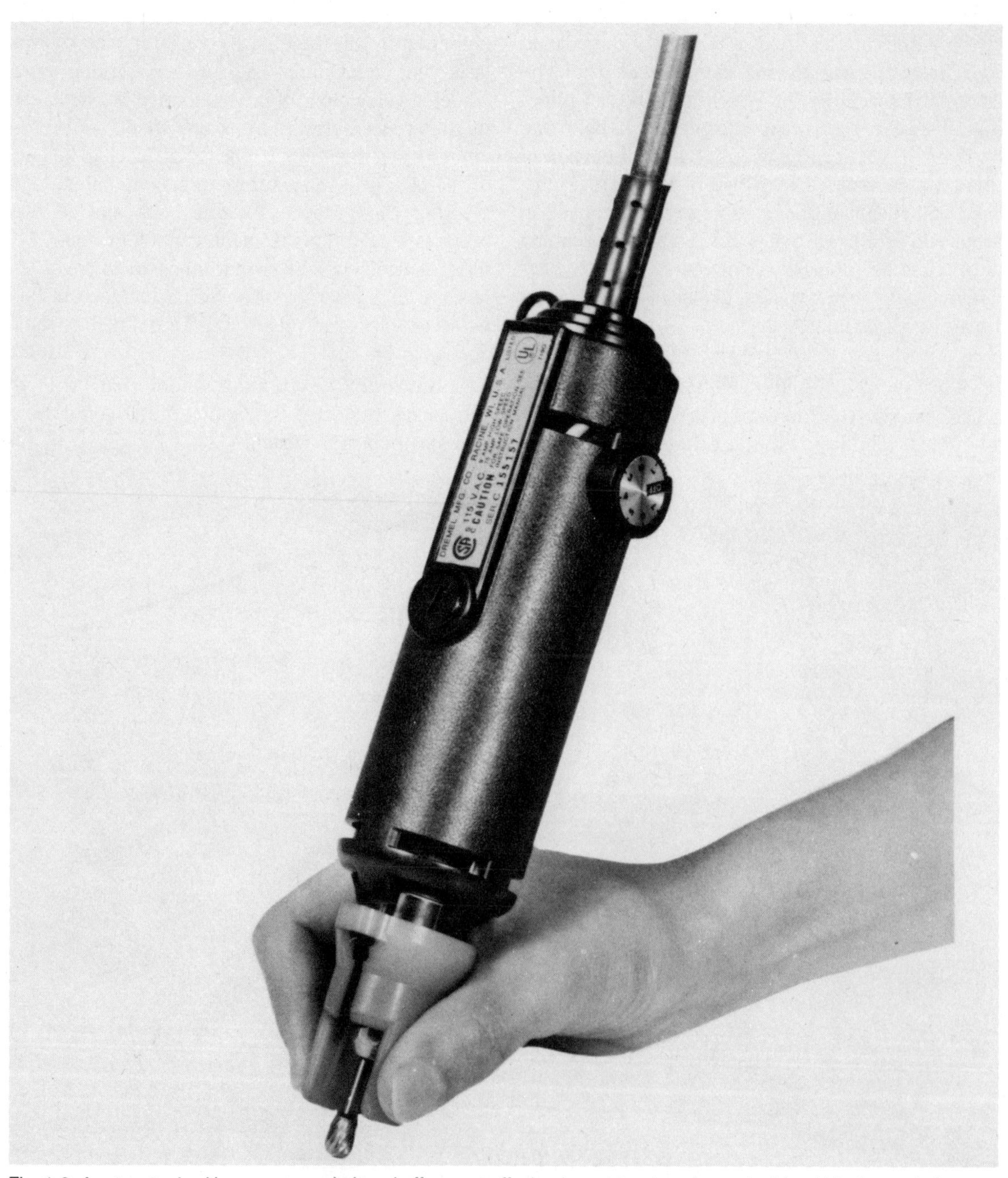

Fig. 1-6. A rotary tool, with accessory grinders, buffers, cut-off wheels and brushes, is a valuable aid in the workplace.

Fig. 1-7. A small bench vise with 2-inch to 4-inch jaws can be bolted to the bench or put away after each use.

Barrel. A hollow cylindrical container, generally of brass, that turns on an arbor and houses the mainspring.

Dial Train. Often called the *motion work*, it is the train of wheels under the dial that moves the hands. It includes the cannon pinion, hour wheel, minute wheel, and pinion.

Escape Wheel. A wheel at the end of the main train that is engaged by the anchor to regulate the clock's running. See Fig. 1-8.

Escapement. A device fitted either with teeth or pins by which the pendulum controls the rate of time keeping. Consisting of the anchor and the escape wheel, it not only regulates the time but gives an impulse to the pendulum. This is accomplished as the anchor releases one tooth at a time on the escape wheel.

Main Train. A series of wheels that conveys the motivating power to the dial train. The barrel (incorporating the mainspring) is the first wheel of the main train.

Pendulum. A body suspended from a fixed point, capable of swinging to and fro freely, used to control the time of the clock.

Pinion. A small wheel, made with six to 12 teeth (called leaves), but not more than 20 teeth. It might be made in one piece or of steel rods in brass endpieces (lantern pinions). They might be an integral part of their arbors as opposed to having been driven on them. A pinion is usually driven by a larger wheel, except in the motion work or dial train and in some electric movements.

Fig. 1-8. An escape wheel with an anchor showing toothed pallets.

Train. A series of meshing wheels and pinions through which power is transmitted from its source (weights, springs, or an electric motor) to the escapement.

Wheel. A circular piece of metal on the circumference of which teeth of various shapes and numbers have been cut.

HOW THE MOVEMENT WORKS

Many, many years ago, my father was given a Seth Thomas mantel clock in a beautiful pine and rosewood veneer case. It was fitted with a silvered dial and a brass movement and had a completely charming chime. He set it on the mantel proudly and we (the children) all marveled at it for a day or two. Without warning, on the third or fourth day, it suddenly went *Uuurrk* and ceased to run. My father decided to fix it.

This promptly gave the rest of the family pause for thought because, while he was not totally inept, his comprehension of things mechanical was nonexistant. Off came the back of the clock and out came the beautiful brass movement (the hands having been destroyed in the process). He stared at it pensively for a moment or two then literally attacked it with a screwdriver.

We gathered about him with our eyes wide with curiosity. Screws came out and little pieces fell on the floor. Suddenly the mainspring let go with an ear-shattering SPROING!!—gashing him deeply in the index finger. Needless to say, we all scattered like flies and watched from a distance as he attempted to staunch the flow of blood with a barrage of epithets and a crumpled handkerchief.

That was the end of the repair session. The movement was shoved back into the case in a most haphazard fashion and it stood on a remote corner for months before disappearing. None of us had dared touch it. Even now, with some knowledge of horology, I wonder what went wrong and—more important—what had made it expire in the first place.

Object Lesson No. 1: The mainspring *must* be let down (made to unwind) slowly before anything else is done. This is dealt with in upcoming chapters.

That sometimes lethal mainspring is but one part in one of four mechanisms in a clock movement. It belongs to the driving mechanism. The other three are the transmitting, the time-indicating, and the controlling mechanisms. As you know by now, the power is stored in the coiled mainspring to which a toothed wheel is attached. The power wends its way through a series of wheels and pinions to the escape wheel.

The controlling mechanism comes into play here in the form of the anchor. This is a device the name of which describes its shape. It is fitted with two teeth or pins, depending upon the type of clock it controls. It rocks on a pivot and the pins thrust themselves between the teeth of the escape wheel alternatingly, restricting it to a definite number of revolutions per minute. This anchor is attached to a crutch (in some cases which controls a pendulum). In other instances, an extension on the anchor controls a balance wheel and a hairspring.

The time indicating mechanism is made up of the dial train, the dial, and the hands. Between the dial and the front plate lies the dial train.

In a simple alarm clock, there are generally four wheels and the escape wheel which carry the power to the dial train. The cannon pinion is friction-fitted to the center wheel to allow for setting the hands.

Historically, all sorts of devices have been suspended between, in front of, and behind the two plates that—with four pillars—hold the clock works together. There are striking, chiming, and alarm systems as well as calendar works and wheels to show the phases of the moon. Some of them will be dealt with in this book at greater length than others, but repair techniques do not differ drastically.

Chapter 2

Clock Kits

THE MANUFACTURE OF KITS TO MAKE ALMOST any item that one could ever want is a huge and profitable business here in America and elsewhere. Automobile enthusiasts, steam engine buffs and airplane devotees, young and old, spend countless hours assembling models of vintage vehicles. These kits vary in a number of aspects: assembly time, degree of difficulty, materials used, detailing and quantity of parts, and finished appearance.

Clocks figure prominently in the realm of kits. Initially, it would appear that the prime inducement would be a finished, working grandfather clock or tall-case clock, proudly displayed for just about one-third the cost of the same timepiece purchased already assembled from a store. The odd factor here is that many people who build one clock frequently do not stop there, but go on building them. This is particularly true in the case of wood-cased clocks. There cannot be a great amount of profit in this venture except, perhaps, for the kit manufacturer. Can the popularity of kit clocks be attributed to the love of wood? The movements are most frequently all assembled requiring nothing more than the fitting of the dial, the hand, and the pendulum. The love of clocks as clocks would appear to be a secondary factor.

A prime example of the disparity in this method of acquiring a clock can be cited here. Early in 1982, the Emperor Clock Company offered what is termed a "Circa 1900 Mantel Clock" in kit form. Included in the package were a solid cherry case kit, a solid brass movement, dial, all hardware and screws, plus easy-to-follow instructions. The sale price from Emperor was $139.50. A photo (Fig. I-3) of the original clock from which this kit was patterned appears in the introduction of this book. These originals turn up repeatedly at flea markets priced around $50, depending upon the condition of the exterior case and whether or not they are in

Fig. 2-1. A 1500 AD mantel kit clock, snap-fit, plastic construction by Lindberg Products, Inc., Skokie, Illinois.

good running order (GRO). One with a badly faded dial and some rust in the mechanism, but working, sold for $25. If "the clock is the thing," which purchase would be the better investment?

Conversely, the number of kits placing little or no importance on the case is increasing. More manufacturers are offering assemblies with no cases at all. These should appeal to the budding horologist and to the dyed-in-the-wool clock lover.

1500 AD MANTEL KIT CLOCK

Lindberg Products Inc., Skokie, Illinois produced a snap-fit plastic construction Mantel Clock kit (Figs. 2-1 and 2-5) early in the 1970s. Although a child could build this model, all the basic rudiments of a simple working timepiece are incorporated. No serious attempt was made to use horological terms in describing or labeling the parts. All wheels are referred to as "gears" and the pins as "wire shafts." It has a true verge and foliot escapement, although the verge arbor is referred to simply as "shaft 24" and the pallets as "metal escapement 35." The verge suspension cord becomes the wag-spindle loop.

These minor discrepancies in nomenclature should not deter you from starting off with this truly inexpensive kit. The manufacturer makes no claim as to accurate timekeeping. The assembly does tick

Fig. 2-2. Parts for 1500 AD mantel clock laid out prior to assembly.

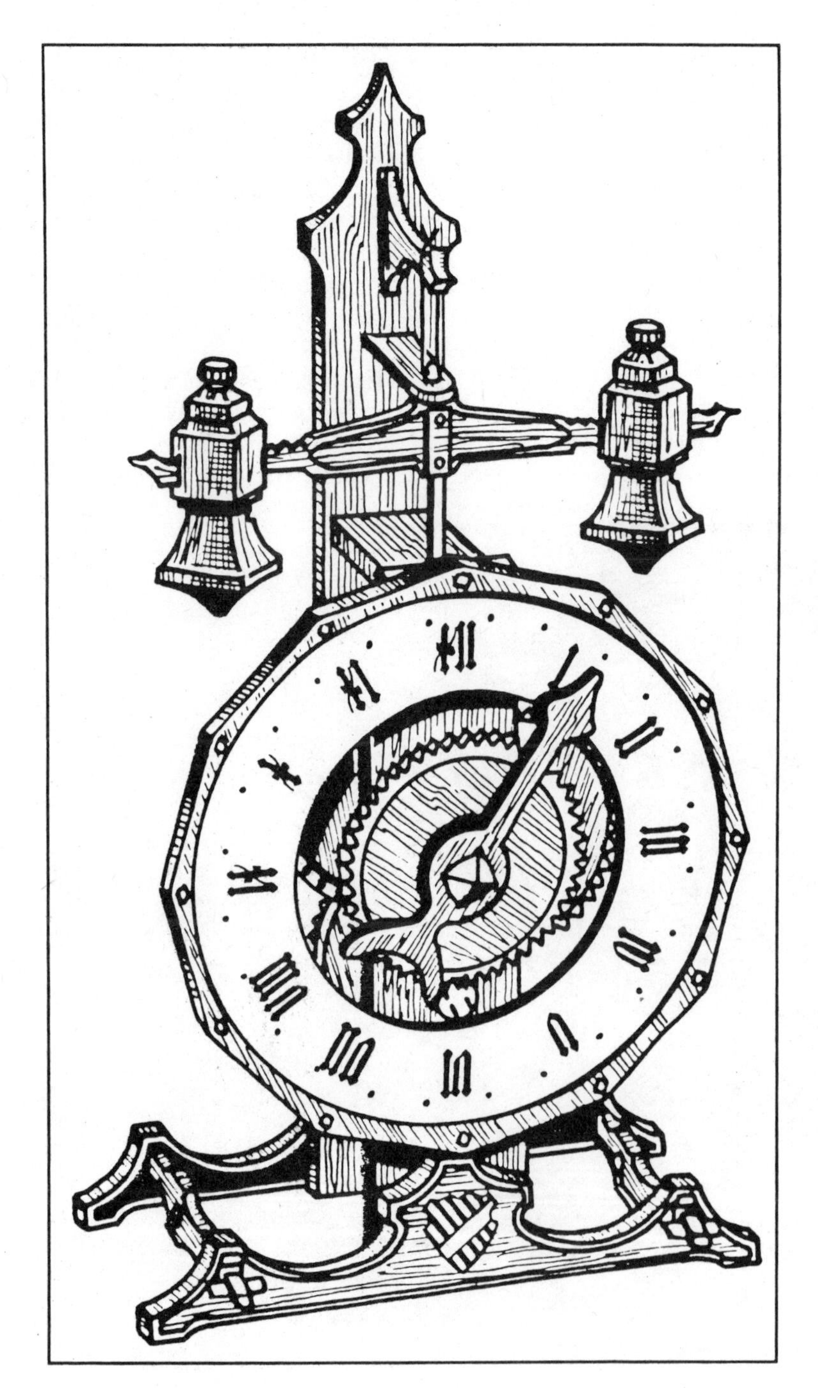

Fig. 2-3. Lindberg drawing of the mantel clock as shown on the assembled plan provided with the kit.

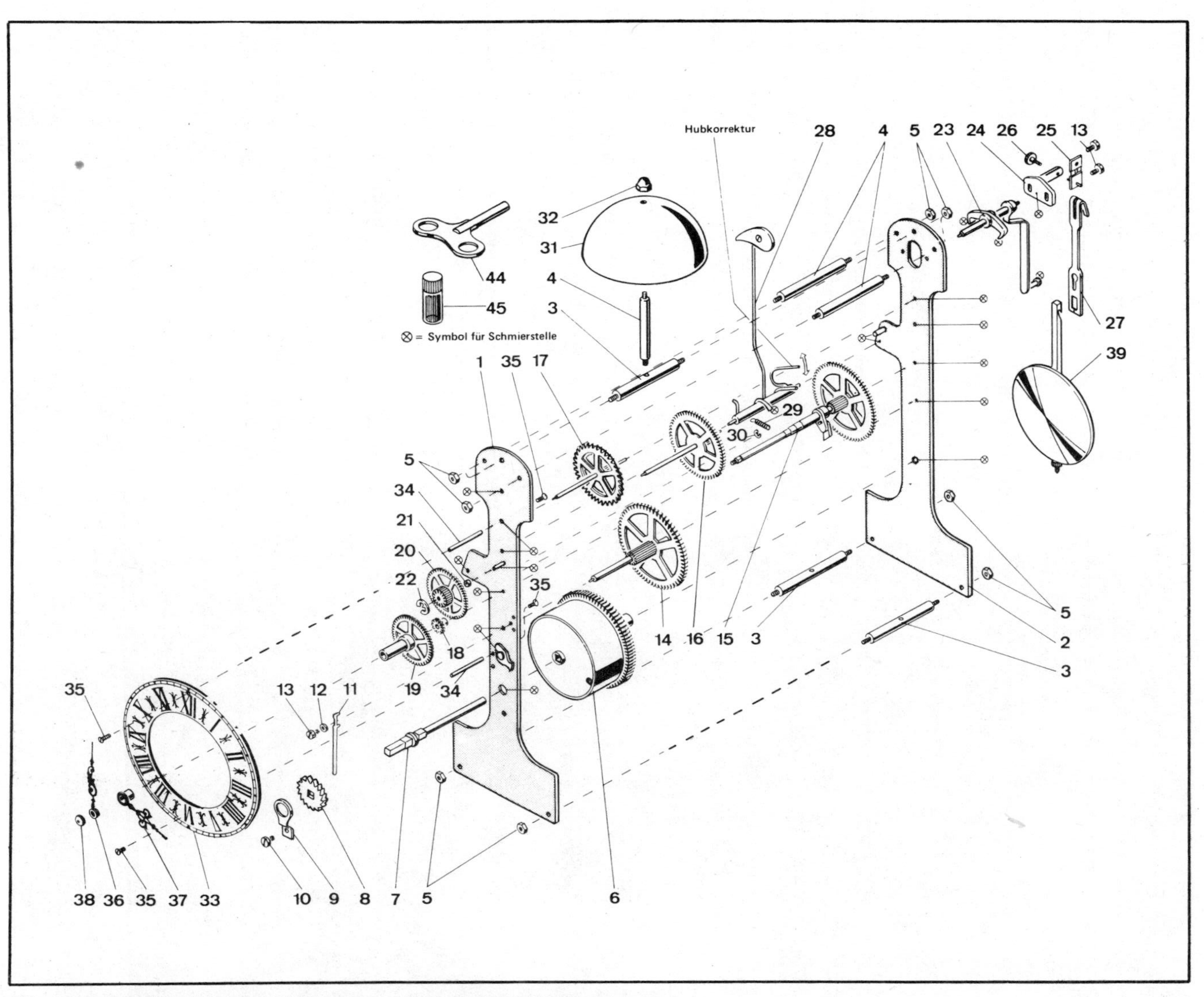

Fig. 2-4. An exploded drawing of Kieninger striking skeleton kit components.

Fig. 2-5. A skeleton kit on completion of assembly.

on and on for hours and an adjustment can be made through the repositioning of the regulator weights. The dial (a decal), with a hint of antiquing, is quite legible; but I wonder whether or not the "4" on the original clock was "IV" or "IIII." The latter was used quite extensively at the time to balance the spread of the "VIII." The little pencil sharpener model shown with the finished model in Fig. 2-1 has the "IIII."

No lubrication is necessary with this model (all the parts are plastic), but care should be taken that the pin height gauge be used to insert the metal pins in the crown wheel or the clock will stop periodically. Care must also be exercised in winding the clock. If the spring is wound past the decal strip indicator provided, the end will come off the storage drum. It is a veritable nuisance to replace. Assembly time for the Lindberg Kit No. 341, even with the careful trimming of all the plastic parts, should be no more than three hours.

Caldwell Industries, Box 170, Luling, Texas 78648, offers two similar kits for weight-driven wall hanging clocks. One has plastic parts and the other has wood parts. At $47.50 the wood kit is almost six times as expensive as the two plastic models. Some builders, nevertheless, might prefer the added authenticity of the wooden wheels.

KIENINGER STRIKING SKELETON KIT

The Kieninger striking skeleton kit first appeared in the Fall 1981 issue of the Westwood Clocks 'N Kits catalog. The movement kit alone was priced at $69.95. An unfinished solid oak base and a glass dome would have cost an additional $26. At about the same time, the Fall-Winter '81 Mason & Sullivan catalog appeared with the same clock in color on its cover, but listed at $97.50 without dome, base or, in their case, a French style, four-sided display case. I succumbed to its charm (and its one hour assembly time) and ordered it at the lower Westwood price.

Skeleton clocks were developed in the early-to-middle eighteenth century. They were obviously an indication of the pride some of the masters took in their work. Great pains were taken to cut away the front and back plate so that the exposed wheel work could be admired. Their popularity rose dramatically in the 1850s and 1860s. The English exhibited some most elaborate clocks of this type patterned after well-known cathedrals (St. Pauls and Lichfield among them). Silas Burnham Terry produced some beautiful examples of skeleton timepieces in America from 1845 on through 1850. One of his pieces is housed in the Old Sturbridge Village Museum today.

Some authorities attribute the popularity of this particular type of clock to the concurrence of its appearance with the then burgeoning industrial age. Could it not be that sheer beauty plus the audacity in the presentation of such extraordinary craftsmanship needs no other leg on which to stand? An upsurge of skeleton clock popularity right now is apparent with a quick perusal of the horological journals, where completely assembled specimens are offered for just under $2300. Elaborate kits retail for as much as $1500. These are special timepieces, indeed, despite the fact that telling time on some of the elaborate specimens is just a bit difficult.

A 12-page booklet came with the kit as well as a 3-page pamphlet (all in German). Following the instructions step-by-step posed no problems and assembly was accomplished with just the four tools suggested: two screwdrivers, a pair of pliers and tweezers. The only adjustment that had to be made was to the hammer. Bending it away from the bell after assembly to allow 1/16 of an inch clearance did the trick. There was but one glitch; the acorn nut provided to mount the bell would not screw on. Close inspection revealed that the internal thread had been made improperly. I dug one, similar in all respects, from my "parts bin" and it worked like a charm. See Fig. 2-6.

The tone of the bell is clear and vibrant with the minute reverberation produced when it is struck continuing for many seconds. Charming!

I had elected not to purchase the dome and circular wooden base, preferring to build my own case so that the timepiece could be hung on a wall. A mockup of the case was fabricated out of cardboard initially and then the dimensions were modified somewhat so that the finished case turned out slightly smaller in width and depth with the height remaining the same. I used mirrored acrylic for the

Fig. 2-6. An acrylic case was hand-made for a completed Kieninger skeleton clock. Mirrored acrylic was used for back, top, and bottom pieces of the case.

back panel, the bottom on which the clock is fastened, and for the back portion of the peaked top. Clear acrylic was used for the sides and door. The top and bottom horizontal pieces were extended to the sides and rear so that square brass pillars could compliment the clock in the case. This proved very effective.

The case was designed to accept the clock with little or no room to spare so that viewing was at its maximum. The piece on which the clock rests, it should be noted, must not be any less than 5 inches wide or the left side of the case, when positioned, will restrict movement of the hammer. For building the rest of the case, consider that the clock is 9½ inches tall, 3⅜ inches deep with a chapter ring measuring 3⅞ inches in diameter.

One of the interesting features of both this clock and the Lindberg kit is that all of the wheels are positioned in a vertical row. Benjamin Franklin is credited with having designed a clock with just such a wheel positioning. Admittedly, these are extremely simple kits. They are offered here for their educational value to the novice, although the skeleton clock is of intrinsic value to anyone with a love for clocks. One final note is that the triangle-shaped bottom portion of the case must be left open so the clock can be bolted down permanently.

More than 25 kit manufacturers or distributors appear in the listing at the end of this book. That is a small indication of how prevalent the interest in clock kits is throughout the nation. Some of them, Mason & Sullivan, Emperor Clock Company, General Time, Heath, and Westwood Clocks 'N Kits, to mention a few, are large, well-established businesses having served the public in excess of 30 years.

ELECTRONIC KIT

Attempting to assemble a quartz clock timer from the Heath Company, Benton Harbor, Michigan 49022, is a matter of switching from wheels, pinions and arbors to resistors, capacitors and diodes! Heathkits are assembled by all sorts of people throughout the country. Yet my experience is that this endeavor is not for the faint-hearted. First of all, you *must* become very proficient with the use of a soldering iron. I accepted the first kit of this type reluctantly, worked at it for a while during spare moments, and then farmed it out to a bright, young man who was "into electronics." He finished the assembly, found that the clock would not work, and finally had to return it to the company for technical consultation.

The kit just mentioned comes with a 57-page manual, an illustration booklet (parts pictorial), and a four-page guide to installing parts, proper soldering, and resistor and capacitor codes. A formidable presentation to say the least!

By the time we had received our second kit from the company, we had indeed become proficient with the soldering iron. It was patently clear, however, that proficiency would take second place to extreme care. With a bit of carelessness one could build a "solder bridge"—actually join two connections by allowing the solder to run between them. The term was totally new to me; previous experience would have prohibited me from forming what is normally referred to as a short.

The trick in building this kit is to follow the instructions explicitly. Check off each step as it is completed. Provision has been made for this. Some of the connections are positioned so close together that a magnifying glass is suggested and actually needed (at least with the quality of *this* writer's vision).

The finished product is a nice, compact automobile clock, and that is what I wanted. Building the kit will not help you solve any of the mysteries of battery-powered LED or LCD clocks because the circuitry must be different. Like the little old watchmaker, you still have to venture in that twilight zone of horology. I was more than satisfied with the results, but I also enjoyed getting back to swinging pendulums, slowly revolving wheels, and even "lethal" mainsprings!

The kits that I have elaborated on here could all be made at the clockmaker's bench. This is especially true because they embody work on the mechanism as opposed to the case. More space and a different setup would be required to tackle a tall-case clock. Even construction of a mantel or wall clock with case would require sanding and eventually varnishing best done far away from numerous exposed clockworks. So the tall-case builder winds up a step removed from the majority of clock enthusiasts for whom this book is intended.

Just check the difference between the tools these builders have to acquire and those cited in the tool section. Freatured prominently in Mason and Sullivan, for example, are cabinet scrapers, pipe clamps, band clamps, etc. These are for kits with precut materials. Further proof lies in the books advertised right beside the clock books in this type of catalog: *Adventures in Wood Finishing, The Fine Art of Cabinetmaking, Woodwork Joints,* among several others.

There is a facet of this type of clock building that applies here. The clock mechanism are referred to as *fit-ups*. They permit the uses of wood and almost any other medium on a fairly small scale.

Chapter 3

How to Use Fit-Ups

THE INTEREST IN CLOCKMAKING AND THE INcrease in the number of people who pursue this craft, hobby or, in some cases, profession, has been accelerated by a significant change in the way clockwork housings are made. When clocks were totally dependent on weights for motivating power, the works was carefully set on a board in addition to being fastened with screws to the sides or back of the clock case. The weight of the total unit had to be distributed as evenly as possible. The gradual use of the coiled mainspring made mounting a little simpler, but it was the ingenious device of a threaded collar around the shaft on which the hands are fitted that brought the possibility of clockmaking or, at least, assembly into every home.

There are a number of ways this is done, and all are actually slight variations on the same theme. A threaded collar is brazed, sweated, or crimped to a hole in the center of the front plate of the clockworks. The shaft assembly for the hands goes through. A slightly larger hole is drilled into the front of the case to accept the threaded collar. The unit is then affixed firmly to the case by a hexagonal nut run up on the collar from the front. This nut, or fastener, is frequently circular in configuration with a groove across the diameter. A special open-ended wrench is used to tighten it; a screwdriver will work if it is used carefully. Any slip here will mar the face of the dial. The hands are added and the job is done.

Often threads are recessed into the clockworks. The threaded collar is then combined with the outside nut that passes through the front of the case and into the clock mechanism. Either method is quick, effective, and can be accomplished by anyone. Except for key wind mechanisms, only one hole has to be drilled through the front of the case. In the case of key winds, a second hole to accept the key has to be drilled, generally about an inch below the one for the hands.

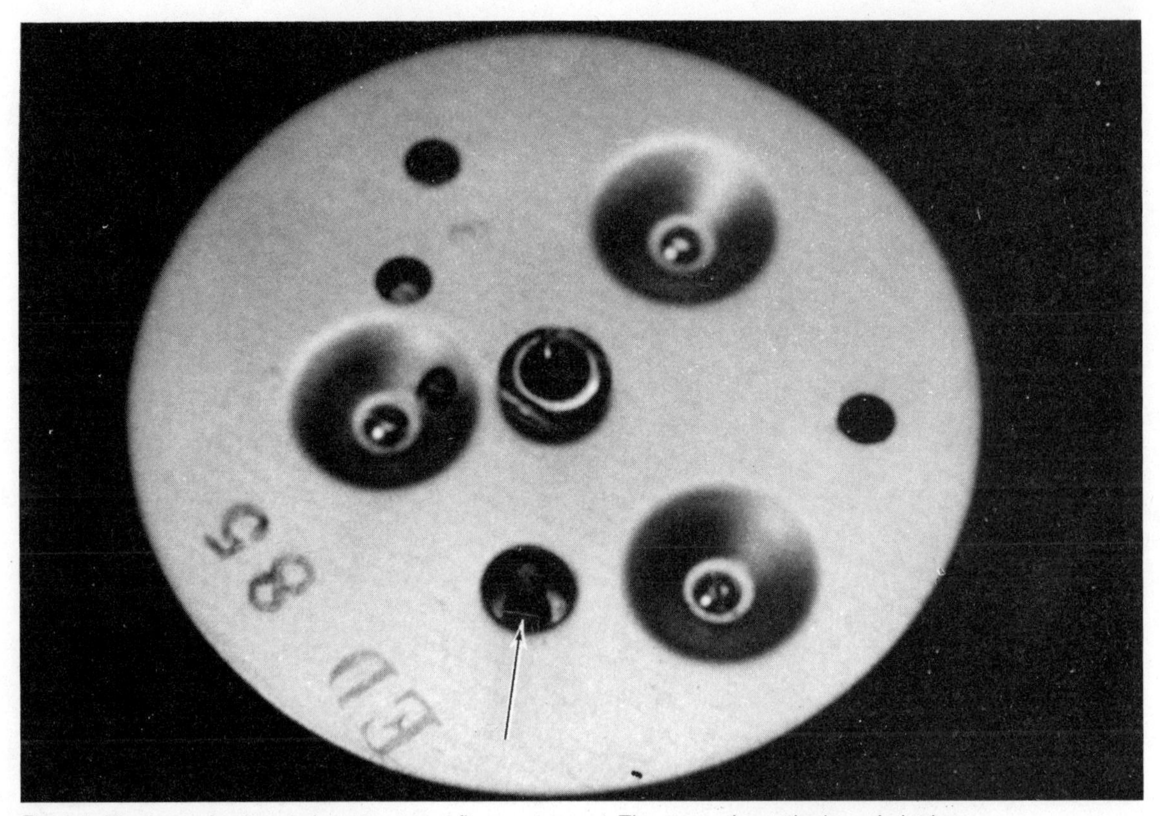

Fig. 3-1. The front of a threaded, center-mount fit-up movement. The arrow shows the key wind arbor.

This now holds true for just about any clockworks irrespective of the motivating power, with the exception of the tall-case clock where weight is still a major factor. This attachment technique is the simpler of two fit-up methods. The alternative, requiring a little more craftsmanship, is discussed later.

The thickness of the plate to which the movement is attached has a direct bearing on the length of the threaded collar. It is advisable, therefore, to gauge the thickness of the plate and select a movement with a shaft long enough to protrude through the drilled hole and to accept the nut. The shafts for the hands always extend far enough out of the collar that all three will clear the dial when fitted.

A variety of movements with threaded center-mount capability are offered through parts catalogs. It is advisable to determine the shaft (collar) length—generally $\frac{3}{8}$ of an inch or $\frac{3}{4}$ of an inch—and the width of the hole in the case or dial before ordering. The length is obtained by measuring from the end of the shaft to the surface of the movement. The thickness of the movement from front to back is also important. Manufacturers now take pride in the fact that they can produce battery-powered movements no thicker than an inch.

Key wind mechanisms generally require a thin dial or case front and utilize the recessed thread mounting technique. The ac-powered movements

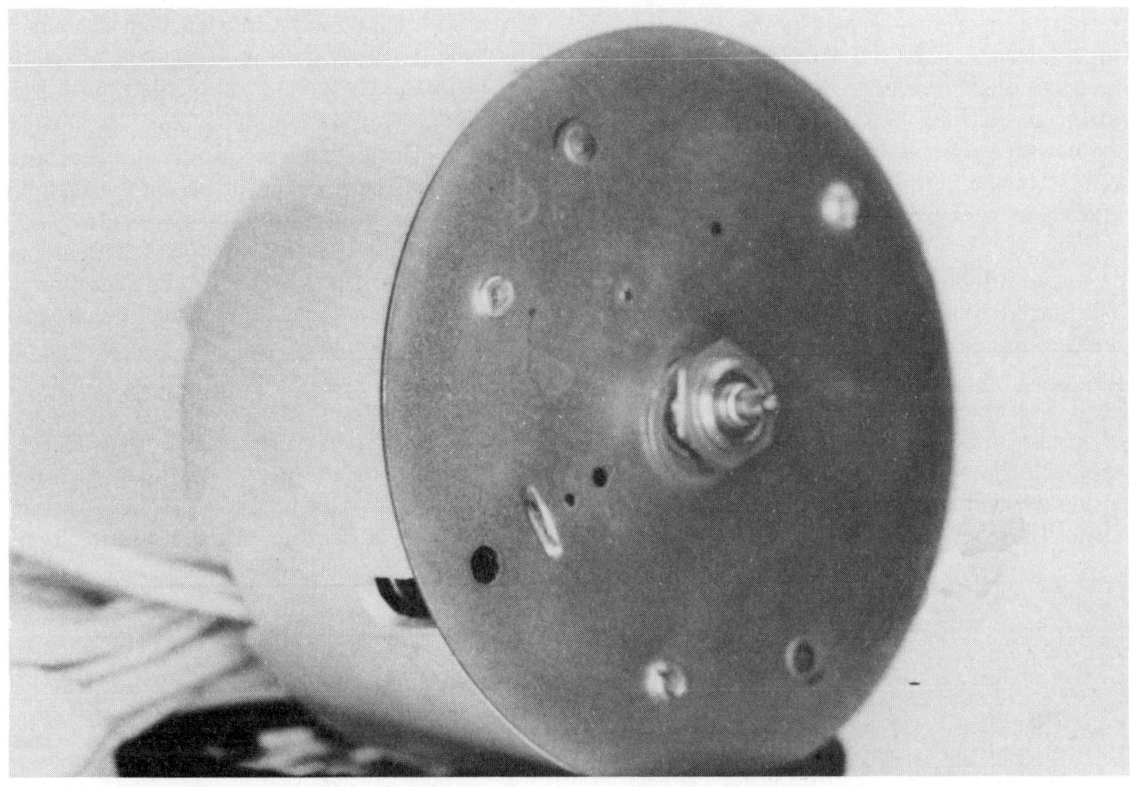

Fig. 3-2. An ac-powered movement with threaded center mount by General Electric.

from General Electric, Westclox, Spartus Corp. and some other manufacturers use the ⅜-inch threaded shaft. The reference here is to movements salvaged from clocks with damaged or obsolete cases. A reliable electric movement made by Lanshire is available with both shaft lengths cited above, and is offered by several of the suppliers listed at the end of this book.

Battery-powered movements are available with shafts in at least three lengths and several diameters depending on the origin of their manufacture (foreign or domestic). General Time Corp. is one of the larger companies offering battery clock kits for persons other than those in the trade.

Many of these movements, in almost mint or excellent working condition, become available through salvage of discarded clocks. People often tire of a style with which they have lived for a while or simply get rid of a good working clock because they are redecorating! This is grist for the clockmaker's mill. It is advisable to exercise caution when cannabilizing these pieces. Some of them fare well to become "collectible" in the not-too-distant future. A French battery movement by Bayard, big and cumbersome by today's standards, should be regarded in this light.

In the series of finished clocks shown in this chapter (Figs. 3-1 through 3-4) an attempt has been made to mate the movement, dial, and hands to an original yet frequently simple hand-made case. A variety of materials was used for these cases or, in some instances, exposed dials. The clocks in this group all have center-threaded mount movements. All originals, they have been named to identify them with the photographs. Many of them have been sold and permission to use them has been granted by the current owner.

CURRIER AND IVES KEY WIND

The nucleus of this clock was a cardboard box top with a Currier and Ives picture bleeding down all of the sides. A snug-fitting liner—a box within a box—was made of scrap wallboard, about a quarter of an inch thick. A thin coat of Elmer's glue was applied to the wood before mounting it permanently inside the cardboard. It was set aside to dry overnight. The exterior was then given several coats of clear varnish and allowed to dry for about 48 hours.

The installation of the clock is simplicity itself. The particular spot for the clock was selected so that it did not detract from the picture. A piece of board was clamped over the picture in the proper spot and a hole large enough to accept the shaft on which the clock hands are fitted was drilled.

An eight-day key wind movement of American manufacture and a dial with Roman numerals was chosen to compliment the Currier and Ives picture. The dial is fitted with a bezel and glass, hinged on the left, that opens away from the clock so that it can be wound weekly.

Hanging the box flush against the wall was accomplished with the use of screw eyes placed about ½ inch in from the back edge. Picture-hanging wire was looped between them and drawn tight

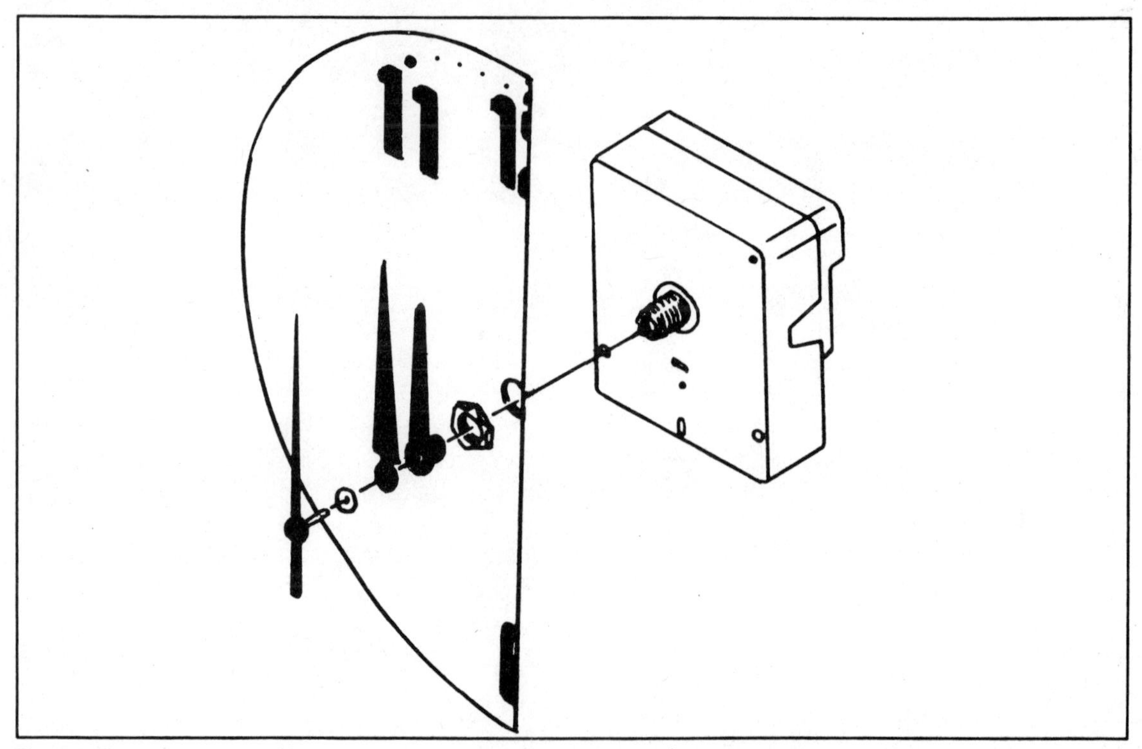

Fig. 3-3. An exploded view of battery-powered, threaded center-mount fit-up movement.

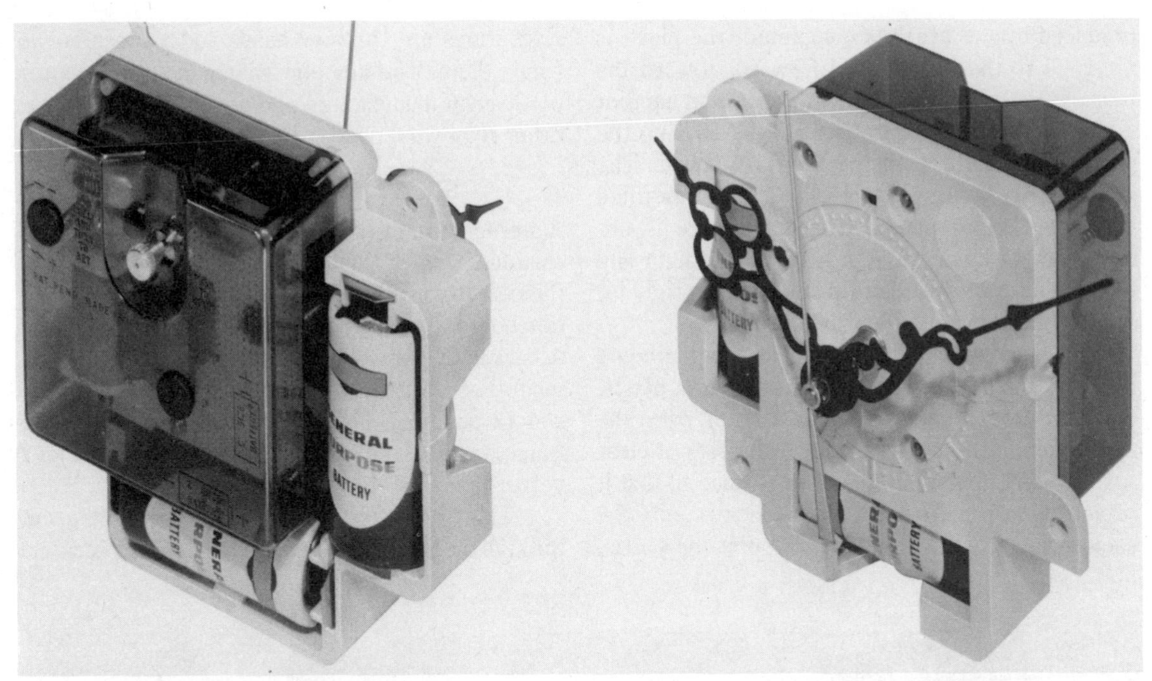

Fig. 3-3A. General Time quartzmatic battery clock kit. Note the positioning of the two batteries.

enough so that the whole box would hang from a strong nail without the wire showing from the top.

FRENCH PORCELAINIZED DIAL ON ACRYLIC

The diamond-shaped porcelainized dial, which became the focusing point for this clock, was rescued from a decrepit timepiece of French origin. The smoked acrylic, measuring 14 × 20 inches, had two bends made in it on the 20-inch dimension with a strip heater. This is a device that works especially well on ¼-inch plastic. The material is laid on the heater (Fig. 6-1) with the area to be bent above the heating element. The heater softens the material after about 10 minutes. It can then be removed from the heater and the plastic can be bent up immediately. It will retain the new configuration if held in that position until the plastic cools completely.

The hole revealing the pendulum was made with a circle cutter. The hole to accept the movement's shaft was drilled, all edges on the plastic were sanded to a matte finish, and all the components were assembled. The dial had a lip so cardboard washers about 2 inches in diameter were fitted between the surface of the acrylic and the back of the dial. This prevented cracking of the beautiful finish on the dial.

KEY WIND WITH BRASS ENGINE TURNED OVAL DIAL

This solid brass dial had black numerals and lines radiating from the center. It was scratched beyond repair, it seemed, so that all traces of the original markings were sanded off the brass. The engine turning applied here is a technique used for hundreds of years to hide fine scratches on metal. The piece is subjected to pressure from a small rotating brush at precise and regular intervals as it is moved from left to right. When a row of the "whorls" produced by the brush is completed, the piece is returned to the left, moved forward (toward the rear of the drill press if one is used), and another series of pressure marks is made by moving the piece to the right until the line is finished. The procedure is continued until the surface is filled with the circular marks. The appearance is enhanced if the lines are kept perfectly straight and the piece is moved the same distance to the right for each application of the whirling brush.

The numerals and batons indicating the hours between them, are adhesive-backed vinyl plastic (available from art and stationary stores). After the application of the numerals, several coats of clear varnish were applied to the brass surface so that it retains its luster. The rest of the clock is quite the same as it was when rescued from someone's attic. Exceptions are the new hands and a brass nut to retain them. The key eight-day wind movement is of German manufacture and required no attention other than winding.

STEREO SPEAKER CONVERSION

A stereo speaker had been housed originally in the case from which this clock was made. The hole in the center was filled with a disc cut from composition board. It was glued in from the rear and braced in back with four short pieces of wood (2 inches by 1 inch). Four ¼-inch holes were drilled at the 3, 6, 9, and 12 position. Then the whole area within the frame was covered with black, flocked Contact for a velvet-like appearance.

The holes for the numerals were then cut through the flocking to accept the hour markers and

Fig. 3-3B. General Time GT 360 battery clock with single dry cell in its own housing below movement.

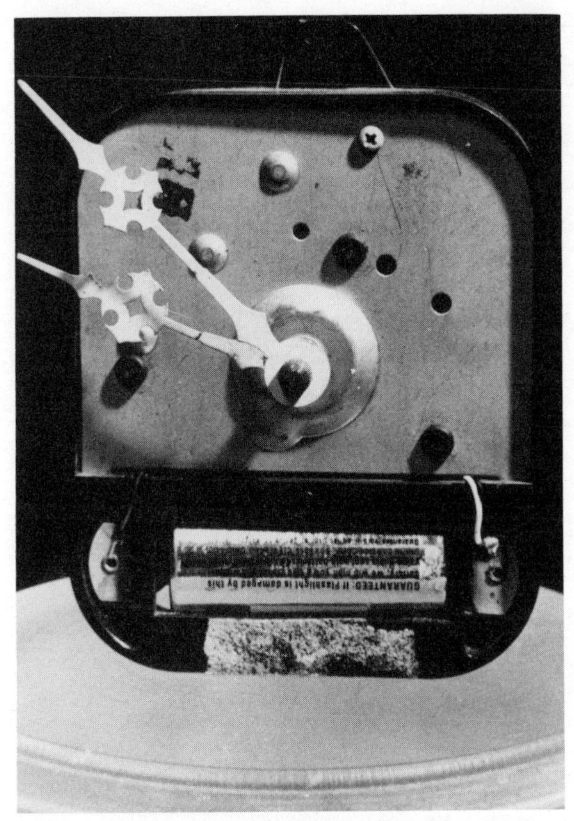

Fig. 3-3C. Early Ingraham battery-powered movement at least twice as large as current movements.

the movement shaft. Drilling through the flocking might have wrinkled it beyond repair. The brass dial was scavenged from a decrepit timepiece, centered on a lathe, and turned while sandpaper was held against it. This is an old clockmaker technique for polishing metal dials. Course-, medium-, and fine-grade paper removed the old numerals and then polished it. It too was sprayed with clear varnish to keep oxidation away. The four brass-finish hour markers were press fitted into the holes and glued.

A small, German battery-powered movement, merchandised by Welby in Chicago, was used. For an elegant effect, serpentine hands were fitted.

NOVELTY PINUP

The dial shown in Fig. 3-16 was found in a box of junk beneath a table in a flea market. A Spartus alternating current movement was used with red, fluorescent hands. A youngster in the house was delighted with it for about six months when, it seems, the rock star went out of vogue!

CANDY DISH DIALS

This winds up being an inexpensive recyling project. A three-tiered candy dish was disassembled, and the hardware was saved for another project. The dishes, clear plastic, had flowers painted on the underside. It was a simple matter to spray paint over the floral arrangement, allow the paint to dry, and fit the movements and hands. Two of them have battery movements and the third is rigged with an ac mechanism. The theory from radio electronics about some shapes being far more effective as sounding boards applies unintentionally here. The battery-powered movement, when hung on a wall, ticks louder than a good-sized alarm clock. The sound of the little motor within the movement winding up to power the train is much more audible than if the movement were encased. Two versions of this example have three-dimensional plastic, adhesive-backed numerals. On the third, the digits are vinyl. These clocks turned out to be very popular for use in kitchens and more and more than a half-dozen of them have been made.

DRUM TABLE CLOCK

The base of this model is a 6-inch wide cylinder of translucent, rippled Plexiglas. The top was cut at an angle and a disc of black plastic was cut to fit the outside diameter. Selection of the battery-powered movement was limited in two aspects: its overall outside dimensions and the mounting of 1.5-volt cell that powered it. Because it was suspended almost parallel to the table but above it, the unit must be one from which the battery will not drop.

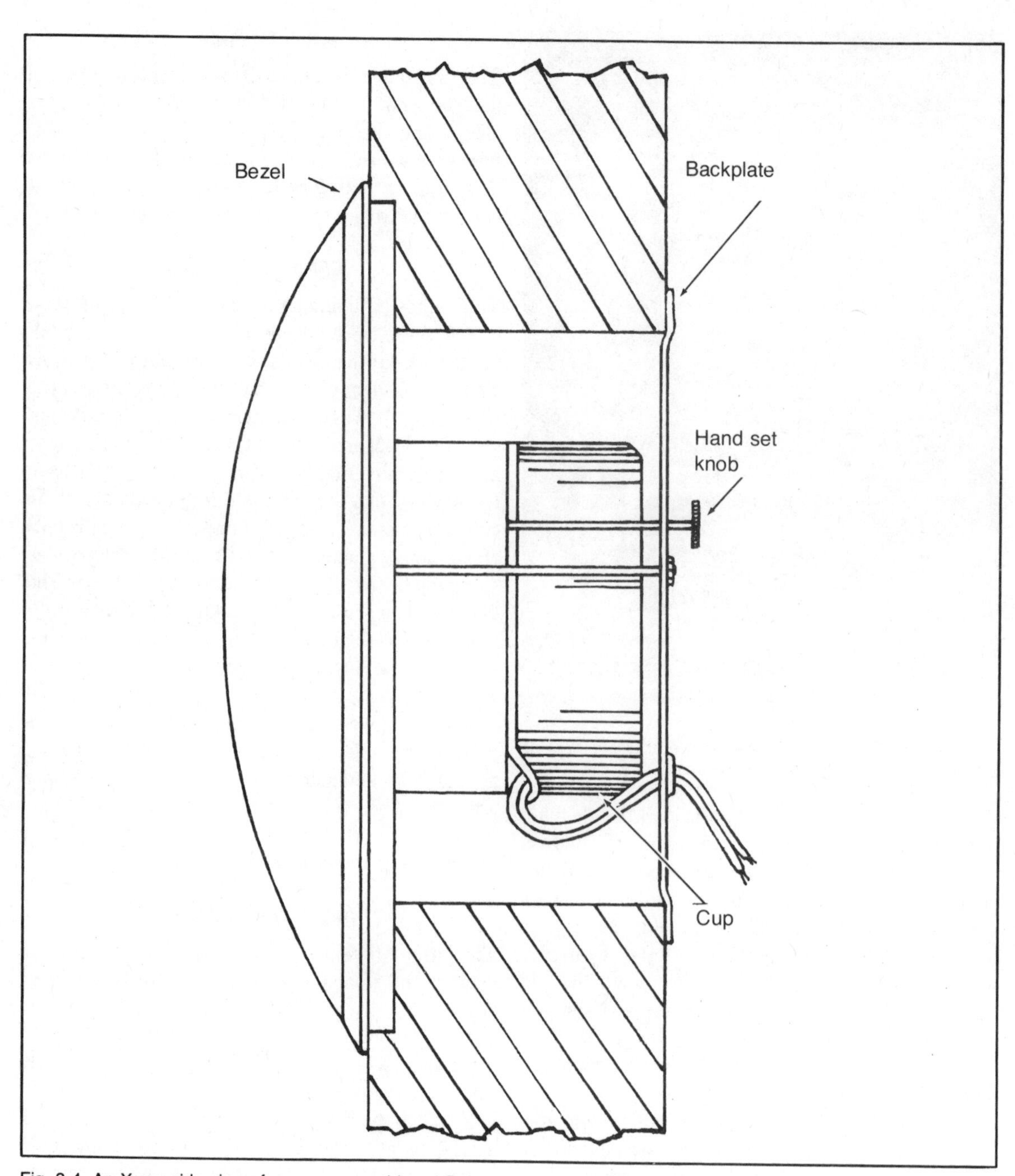

Fig. 3-4. An X-ray side view of an ac-powered bezel fit-up.

Fig. 3-5. Currier and Ives key wind fit-up (front view).

Units are now being manufactured that make provision for the little battery box or holder to be moved to the side or mounted elsewhere than right on the unit for a better fit in tight places.

SQUARE DESK UNIT

A desk unit can be made up very quickly if you are short of Christmas gifts. A square of brown Plexiglas is drilled in the center for the movement mounting. Depressions for what would be the baton positions are made with a drill bit, and then painted in with brass or gold paint. The outer edge, in lieu of being sanded to a matte or mirror finish, is covered with metallic tape all around. The numerals are self-adhesive vinyl. The circular ac-powered movement is fitted with a brass-finish wire stand that holds it at the angle shown. The line cord is hidden behind it.

TRADITIONAL OPEN-FACE MANTEL CLOCK

The wooden case for this mantel clock was acquired at low cost from a retired clock-case maker. Inside of the frame was lined with red flocked Contact. The brass dial, in fairly good condition, was filched from an old electric Seth Thomas clock (note lack of key hole). A reliable General Electric Telechron movement was fitted, and the whole was finished off with flat brass hands. There were no problems with assembly, yet a finished product was obtained that is striking in appearance. See Fig. 3-20.

ACRYLIC AND WOOD DECORATOR CLOCK

This piece is quite similar to the French porcelainized-dial version with the exception that wood was added to the sides after the acrylic was bent back on the strip heater. The dial and ring are a single unit made of aluminum. The numerals and

Fig. 3-6. Currier and Ives key wind fit-up (3/4 view).

batons are vinyl and the hands are aluminum spray painted flat black. The battery-powered pendulum movement is Japanese.

WOODEN TREE LIMB DISC

A wooden souvenir ashtray was reversed to produce the effect shown in Fig. 3-22. The bottom (now the dial) was sanded perfectly smooth to accent the rings. Turning the ashtray over left a hollowed-out area into which a Japanese battery-powered movement employing a single D-cell was fitted. The outer surface of the wood, including the bark, was given numerous coats of Polyurethane. Numerals and dots are adhesive-backed vinyl. Any sound from the movement, incidentally, is much more subdued than in the case of the candy dish unit described previously.

PRINTER'S PLATE WALL CLOCK

This piece attempted to capitalize on one facet of the nostalgia craze. When letterpress printer's plates were being snatched up all over the country, I acquired several and made this clock upon request. The plate was cleaned with a solvent to remove all traces of ink. It was drilled in the center. The board on which it was to be mounted had a hole cut out large enough to accept a small battery-powered movement. Wheels and pinions from our "parts bin" were selected to serve as hour markers. They were cleaned, polished, and then mounted with brass upholsterer's nails. The back of the board was lined with felt and a hanger was added. The whole front had been varnished to preserve the overall appearance.

Pieces like this one always generate curiosity because the printing is reversed. In many cases, the advertiser for whom the plate was made has gone out of business. See Fig. 3-23. If the name here sounds familiar it is because Oster was a well-known manufacturer of barber tools and other cutlery items.

WALL CLOCK WITH RECESSED MOUNTING

This is a conventional, battery-powered fit-up with the exception that the movement was recessed into a wall. The painted aluminum dial is one left over when the Gibralter Clock Company of Jersey City, New Jersey went out of business. It was adhered to an ⅛-inch-thick square of acrylic drilled at the four corners. The whole unit was then mounted in a bathroom with the movement just inside the wall in a hole cut to fit it (same technique as the printer's plate clock). Replacement of the battery is accomplished by reaching one's hand in through the medicine cabinet; that's quicker than removing the four mounting screws!

LEATHER-FACED WALL CLOCK

Scraps of four different kinds of leather, the same thickness, were selected and cut to size (to fit a 10-inch square). They were carefully glued to a piece of ¼-inch plywood. The numerals and dots were tooled into the leather, and then painted with gold paint. The frame was mounted and a General Electric ac movement was fitted. The distance from front to back after assembly was 3 inches. That makes this one appear to stand out from the wall more than most of the others presented here.

Fig. 3-7. Currier and Ives key wind fit-up (back view).

Fig. 3-8. French porcelainized dial on acrylic.

The clocks that follow are built using the threaded center-mount technique, but have cases built from scratch or have some special feature about them. All except one are fabricated of some form of plastic. In essence, these cases are catching up with the on-going changes in movements. Many movements are made almost entirely of plastic today.

BLACK BEAUTY

The black beauty is one of the very first battery and pendulum-powered movements with which I experimented. The movement was acquired with no pendulum and a warped, aged wooden case. The length of the pendulum needed was calculated (data on this appears farther on) and then the case was built to house the movement for which a new pendulum was crafted. The chapter ring used has a brushed silver finish with black markings.

A disc of black acrylic was cut with a circle cutter, then turned on a lathe, and coarse sandpaper was applied to the surface. This produced lines similar to those on a record. The case was built with the front piece cut to accept the circle with a close fit. The back of the case was fabricated of mirrored acrylic, with the piece being cut off slightly taller than the front piece. The side pieces went all the way up in back of the disc to support it and the movement.

Discs of the same black acrylic (1/8 of an inch thick) were cut about 2 inches in diameter and edges were beveled. They were cemented with acrylic cement and then the solid disc was drilled

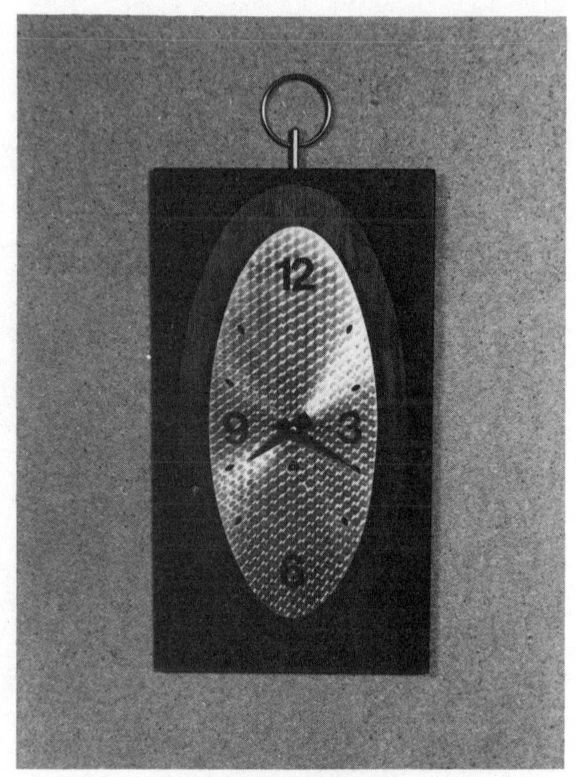

Fig. 3-9. Key wind with brass engine-turned oval dial.

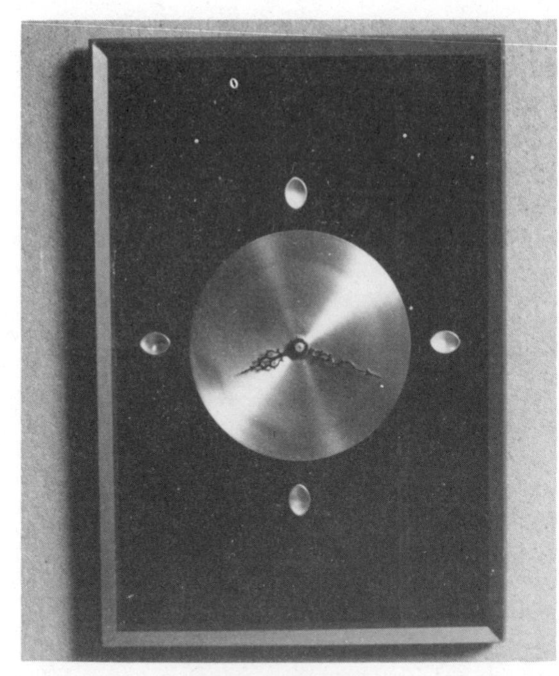

Fig. 3-10. Stereo speaker conversion.

and threaded to accept a 6-×-32 machine bolt. A quarter-inch square plastic rod was fabricated for the pendulum rod and the bottom end of this was drilled and threaded. The top end was cut to fit the end of the crutch.

The bolt was rounded off on both ends so that it would screw into the pendulum bob and into the rod. Two nuts were put on: one to lock up against the rod and the other to prevent the bob from rotating. The length of the bolt was such that the bob could be screwed up to regulate the pendulum. Plain brass hands are fitted so that the clock did not have an overly ornate appearance. The lower front portion of the case was drilled to accept a brass-finished bezel without glass. The mirrored back picks up light through this hole and silhouettes the pendulum as it moves back and forth. In spite of the heavy use of black, the time is very easy to read on this clock.

CUBE WITH FLOATING MOVEMENT

The dial or plate for this timepiece was cut to fit

Fig. 3-11. Original speaker housing with center filled in.

Fig. 3-12. Speaker housing with black flocked contact, brass, disc, and serpentine hands added.

diagonally across a 4-inch plastic cube. The outside edges had to be beveled front and back so that it would slide up into the inverted cube. An oval of black Contact hides the little battery-powered mechanism that would not fit into the cube without 1-inch slots being cut into the back faces of the cube. Vinyl numerals—just the 3, 6, 9 and 12—were affixed to the dial and the whole unit appears to be floating in the cube. In spite of the miniature size of the movement, the tick it produces must be magnified by the walls of the cube because it is louder than many of the other clocks I have built. The circular plastic platform shown in Fig. 3-29 is not a part of the clock.

NOVELTY CYLINDER DISC WITH PRISMATIC CENTER

The exterior of this clock case came from a blue plastic cylinder 6 inches in diameter and 4 inches wide. It was cut to 2 inches in width and the remaining portion of the cylinder was cut into five pieces. These pieces were turned so that the curves flowed inward while producing a star effect. They were glued in this position. The space remaining within the star was so small that only a concerted search turned up an ac-powered movement that would fit. It was hidden behind a disc of prismatic tape. Within the top point of the star is a neon bulb of the type used to illuminate some clock dials at night.

The plastic circular disc fitted to the front that holds the movement has impressions drilled into it from the rear that are filled with white acrylic paint. Short lengths of rigid plastic tubes are cemented to the bottom to serve as stabilizing legs. The U-shaped platform that appears in Fig. 3-30 is not a part of the clock.

DISC AND DRUM WITH LIGHT IN BASE

Here a large (10-inch) disc, cut with a beveled edge, served as the dial and movement support. It was "engraved" with lines and dots from the rear and the edge was sanded to a matte finish to pick up the light piped from the incandescent light in the translucent white base. The base is actually half a drum in shape made of a piece of curved white plastic with front and back pieces fitted to the curve and cemented on. The disc is cemented into a slot in the top of the drum. The white area in the center is Contact glued from the rear. The fit-up is a GE alternating-current movement.

SWIRLS ON METAL

The novelty here is in the way the paint was applied

Fig. 3-13. Movement, housing, hands, and dial for stereo speaker conversion before restoration.

to the metal disc of which this clock "case" is made. See Fig. 3-32. Several coats of white primer were applied to the metal, and then additional coats of glossy white enamel were added. When thoroughly dry (after about 48 hours) the areas where the numerals appear was masked off with masking tape. A technique called "dunk-it" was used to produce the swirled effect. Krylon paint was sprayed on the surface of a bucket of water, stirred gently, and then the disc was immersed through the paint down under the water. It was held there until the paint "skinned" on the surface and was removed. The disc was then removed from the bath and carefully dried.

When the surface was thoroughly dry, the numbers were applied and the whole surface was treated to several coats of varnish. An ac-powered movement was fitted and brass hands were applied.

ENGRAVED DIAMOND-SHAPED DIAL

This clear acrylic dial (¼-inch thick) on an illuminated base is personalized with an Old English L engraved in the center. The base, 4½ inches square by 3 inches high is made of red, translucent acrylic.

Fig. 3-14. Closeup of movement for stereo speaker conversion.

Built into it is a standard lamp socket with a 7 ½-watt bulb that can be switched off at will from the outside. The dots are drill impressions done on the rear of the dial. The letter is engraved in the center with the aid of an impact-type engraving tool. This tool is manufactured under the name of the Electro-Stylus and is fitted with a fine diamond point.

An Ingraham ac unit was installed because electricity is needed for the light in the base and to illuminate the dial. This metal place on the movement was covered with deep red Contact that matches the base in color and provides a background for the engraved L. Versions of the clock have been made with the bases in different colors with emblems such as the VW Rabbit engraved in the center.

ROTATING RINGS UNDER A DOME

This timepiece begins as a fit-up, but winds up being a pure novelty! The dial reposes face up on a drum-shaped pedestal with plastic rings serving as hands. Telling the time on it is not that difficult when you determine that the outside ring is the minute hand, the inner white plastic ring serves to "turn" off the hours, and the smallest circle (of mirrored acrylic) revolves on the shaft normally fitted with a sweep second hand. Black arrows are fastened to one side of the minute and hour ring to reduce whatever confusion might ensue.

Fig. 3-15. Closeup view of movement (rotated) for stereo speaker conversion.

The example shown in Fig. 3-35, without the dome, indicates "20 minutes to 6". The black arrows are duplicated on the inner surfaces of the white rings so that the clock does not have to be moved to tell the time. An ac movement, by Wm. L. Gilbert Company, Winsted, Connecticut was selected to power this specimen solely because the unit, including the back cover, was round and fit up into the outside cylinder without modification.

Fig. 3-16. Novelty pinup approximately 9 inches in diameter.

FIT-UPS USING INSERT MOVEMENT

A variety of movements fall into this category. There are electric-clock (ac) examples, 30-hour key winds, even battery-powered quartz movements. Installation is a little more difficult here

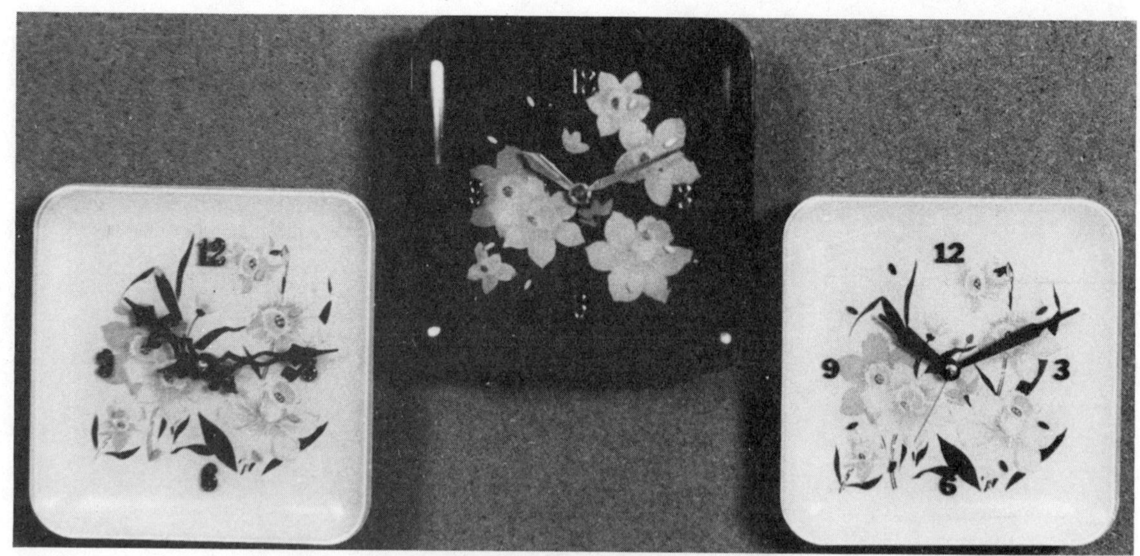

Fig. 3-17. Candy dish dials make a simple recycling project.

Fig. 3-18. Drum table clock from cylinder of translucent, rippled Plexiglas. Owner: Marjorie Coggins.

Fig. 3-19. Square desk unit with self-adhesive numerals and brass hands on brown acrylic dial.

because the front of the case or the stand must have a hole in it, ranging in diameter from approximately 2 inches to a bit more than 5 inches (depending upon the unit used). Prominent in this field are the units made by the Lanshire Clock and Instrument Corp., Chicago, Illinois. These can be purchased new from clock movement suppliers.

Anyone with the inclination and time to go scavenging can turn up serviceable units—requiring little more than cleaning—made by Sessions, Seth Thomas, Gilbert and United for a dollar or two. Cutting a 5-inch diameter (or smaller) hole presents no serious problem.

Another difficulty that might be encountered lies in the measurement (front to back) of the case to be used. These units generally have a plate, with two holes in it, that is fitted from the rear over two threaded studs and tightened with two nuts. Most of the plates are flat. If the case is too thick, the studs will not protrude enough so that the nuts can be fitted. Mounting in such a case is accomplished by the use of a plate that is dished or recessed. A lip rests on the back of the case while the recessed portion reaches down toward the mechanism deep enough so that the rods will come through. Fabricating the plate out of metal might prove extremely difficult, but it can be done in acrylic. The clocks that follow embody several solutions.

Fig. 3-20. Traditional openface mantel clock. GE Telechron movement with brass hands and dial on red flocked Contact.

Fig. 3-21. Acrylic and wood decorator clock with aluminum dial and ring in one unit. Japanese battery-powered pendulum movement was used.

Fig. 3-22. Wooden tree limb disc made from souvenir ashtray reversed to accept Japanese battery movement.

ARCH IN ACRYLIC

Three pieces of acrylic were literally fused together to make this small (9 inches high) timepiece

Fig. 3-23. Printer's plate wall clock on wooden plaque with wheels and pinions denoting hours.

with its elegant crystal-like appearance. The center block (just under the engraving) was embedded in a precisely routed-out area on the bottom piece and fixed there permanently with acrylic cement.

A 3⅜-inch hole was cut first for the movement in the top piece. Then it was rounded off to match the hole's curvature. The outer edge was polished to a matte finish after the engraving (which is always better when done on the back) was finished. The design was taken from an old East Indian wooden textile die.

Fig. 3-24. Reverse side of printer's plate clock showing center cutout to house battery movement.

Fig. 3-25. View with all components for printer's plate clock prior to assembly.

Because there was no case with sides and back to support the movement, a plastic cylinder with a 3½-inch inside diameter, 1⅝ inches deep was used. A thin, circular plastic plate cut to fit was cemented to one end and then drilled with three holes—two for the studs and a larger one for the hand set. One additional hole was drilled in the side to allow passage of the line cord for the Lanshire ac movement. One feature of these Lanshire movements (new or used) that we've come across to date is that all the wheels, with the exception of the fiber one next to the motor pinion, are black metal rather than brass or uncoated steel.

Fig. 3-26. A wall clock with recessed mounting. A hole was cut in the wallboard to accept the battery movement. A thin acrylic plate holding the dial is screwed to the wall.

Fig. 3-27. A leather-faced wall clock. Various kinds of leather matched in thickness are mounted on 1/4-inch plywood, 10 inches square. Owner: Richard Hillman.

Fig. 3-28. A battery-powered pendulum fit-up with a brushed silver chaper ring on black acrylic.

DESK CLOCK WITH MIRRORED ACRYLIC SURROUND

This is another approach to the treatment of a six inch diameter cylinder. This one is of the same material as used in the Novelty Cylinder with Prismatic Disc clock. It was sanded thoroughly to give the surface a "bite" and then dipped as described previously (Swirls on Metal). The two-tiered base received the same treatment—all in beige, brown and white. A disc of mirrored ¼-inch acrylic was cut to fit and this, in turn, had the center removed to accept an ac movement by the Sessions Clock Company. Mounting here was with a fabricated cup snubbed up tight to prevent the unit from turning when the hands were set.

LARGE-DIAMETER FIT-UP

The United Clock Company, Brooklyn, New York made these movements (Fig. 3-38) and installed them in a variety of traditional and novel cases. Such was their mass-produced volume that countless numbers of these clocks, generally in excellent working condition, wind up in flea markets, rummage sales, and garage sales with price tags rarely in excess of one or two dollars. They were held in the case with a large metal plate and two studs with nuts. This mounting method applied to solid materials (about 1½ inches thick) through which a hole had been cut. When the facia was thinner, a large metallic cup was used.

The marble stand and the United movement (Fig. 3-39) were combined solely because the hole in the marble was just the right size. All records of what was originally housed in this stand have been lost. It might well have been used by any of a number of clock manufacturers because the 6-inch diameter of the movement is not an unusual one especially for wall or mantel clocks.

The original plate that came with this movement was missing and would not have worked here, in any event, because the marble stand was about ⅝ of an inch thicker from front to back than the initial case. Scrap plastic was used to "cobble up" a plate with a recessed cup attached to it (which sufficed).

Surprisingly, the finished item was not intended for resale, but more to use the marble stand that had been lying around for months. Nostalgia enthusiasts seem to gravitate toward it, making it an unexpected conversation piece.

CLOCK WITH SAND-PAINTED FACE

The mechanical aspects of this assembly are quite similar to a number of the projects explained previously. A Japanese battery-powered movement was bolted up to a 12-inch square of composition board that had been painted with sand in the following manner.

Colored aquarium sand (purchased from art or hobby shops) is dusted on to the surface of the board coated with a mixture of Elmer's Glue and water. Three ounces of the glue is diluted with 1 ounce of water. The surface, in some cases, has to be primed with a light-color paint to prevent light sand colors being degraded by a dark background. The sections are separated by marking them off with pencil. The

Fig. 3-29. A cube with a "floating" movement. The dial fits diagonally across a 4-inch plastic cube.

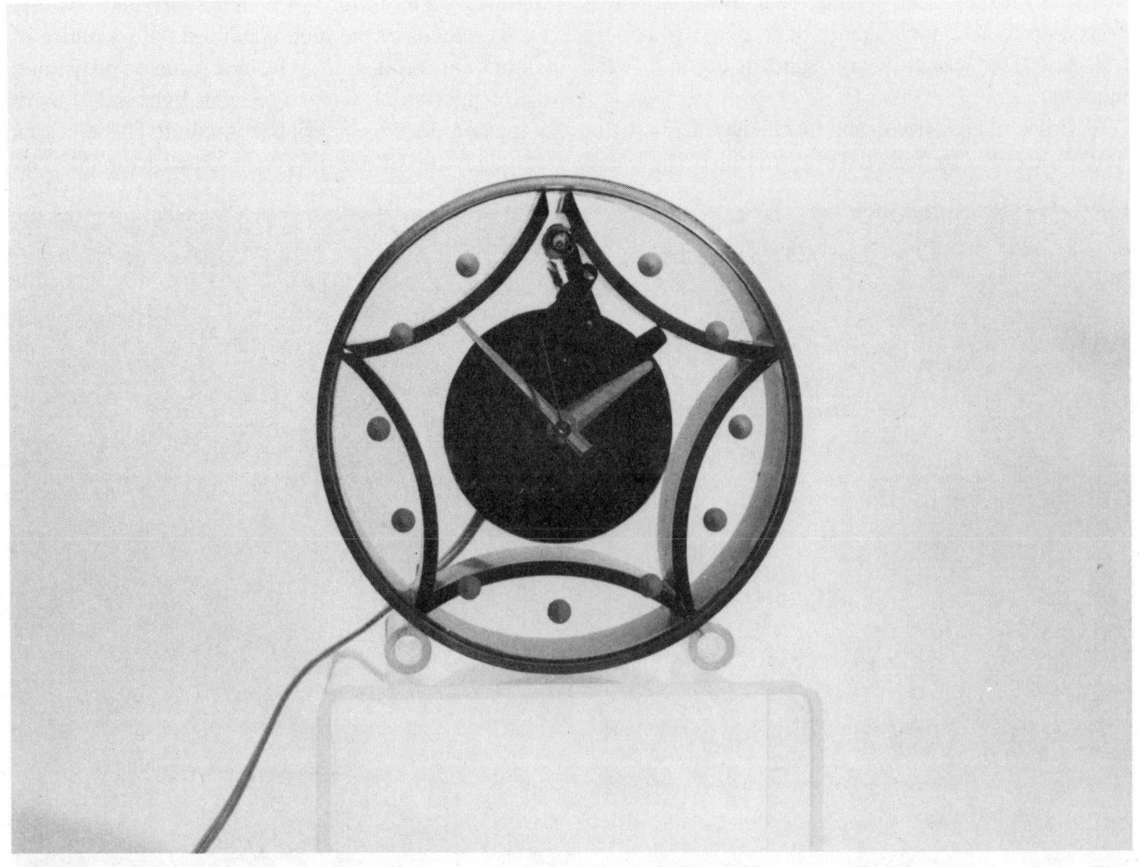

Fig. 3-30. Novelty cylinder disc with prismatic center. The star and the exterior circle were cut from one 6-inch cylinder. The finished depth is 2 inches. The U-shaped acrylic stand is not a part of this clock.

area in the center is masked off with a circle of adhesive-backed Contact.

The colored sand is applied to one area at a time by pouring it gently from a small paper cup. Keep the cup moving as the sand is applied. Allow the sand to dry for about 10 minutes and then pour off the excess on to a sheet of newspaper. It can be poured back into the cup for further use. Now allow the sand to dry for a full half hour before applying a second coat. Two coats are mandatory; three may be needed in some instances.

Certain light-colored sands will not accept the second application of glue. When this happens, spray the whole surface with Krylon Matte Finish or with varnish and set it aside for 15 minutes. Proceed as before once the surface is dry.

Work with one color sand at a time. Apply the glue right to the edge of the adjoining color, but do not overlap. The numerals on this face had been applied prior to the sand treatment. The brass finish disc was placed carefully in the center only after the whole surface had dried thoroughly. Fitting the

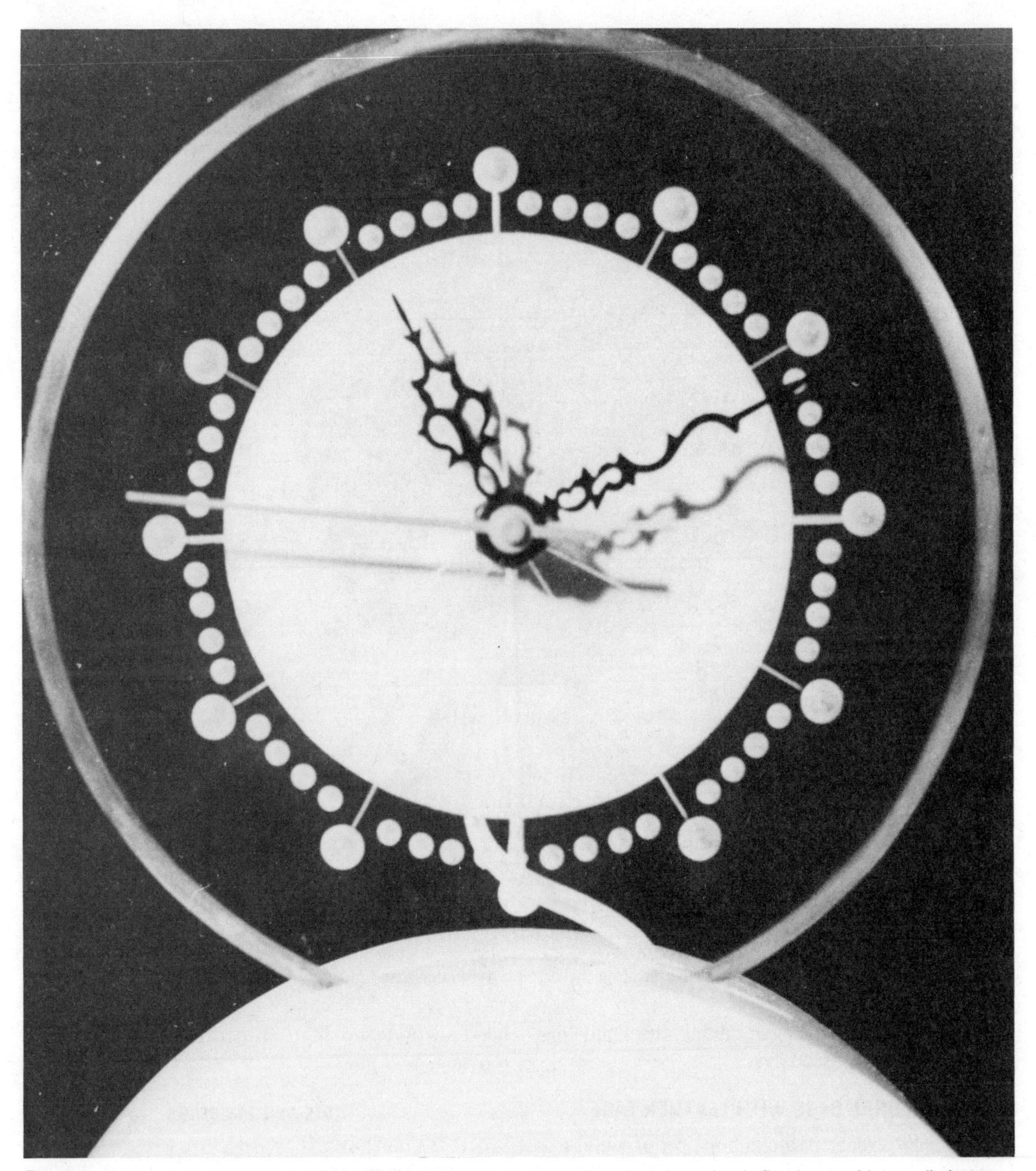

Fig. 3-31. A disc and drum with a light in the base. A 10-inch disc with a beveled edge is fitted to a white acrylic base.

Fig. 3-32. Swirls on metal. A feature of this fit-up is the dipping technique used to apply the paint.

movement with its nut and adding the hands has been explained previously.

HEXAGONAL CASE WITH LEATHER FACE

This simple project is a combination of wood and leather. Most of the materials will be right at hand—wood and leather scraps, quarter-round molding, and glue.

Tools and Materials

3/8″ plywood: 6″ × 16″
1/4″ round, 3′

Fig. 3-33. Engraved diamond-shaped dial. An Old English L was applied with an impact-type engraving tool.

Leather (scrap), 6 pieces (approx. 4″ square)
Drill
Drill bits, 3/8″
Coping Saw
File
Cardboard
Compass
Ruler
Protractor
Sander
Battery clock mechanism
Elmer's Glues: Carpenter's Wood, Wood Filler, Contact Cement, Glue-All

Fig. 3-34. Rotating rings under a dome. The movement is fitted horizontally in a circular drum base. The hands are plastic rings fitted to normal collars.

Fig. 3-35. A rotating ring clock with the dome removed.

X-Acto knife, #11
Adhesive-backed vinyl numerals, ½″
D-cell battery
Green felt, 3″ × 6″
Krylon Wood Primer and Krylon Interior/Exterior Enamel

Procedure

Cut the three base pieces from the ⅛-inch plywood stock as follows: 3″ × 6″, 2½″ × 5½″ and 2″ × 5″. Take the smallest of the three and drill a hole, centered, 1 inch from each end. Then use the coping saw to make two cuts from one hole to the other and the waste piece will drop out of the slot.

Set the base pieces aside and prepare a cardboard template for the face. Draw a 6-inch circle flush with the top edge of a piece of cardboard the same size as the faceplate (6 × 6⅜ inches). Draw six lines radiating out from the center of the circle, 60 degrees apart, and allow them to intersect the circumference of the circle. A little protractor is just right for this step. If a protractor is not right at hand, just draw a line across the center of the circle and mark off 3-inch increments around the circumference. The extra ⅜ of an inch at the bottom

Fig. 3-36. An arch in acrylic. Fusion of three pieces of acrylic house the inserted movement. The clock is 9 inches high. Owner: Mr. & Mrs. James McMeen.

Fig. 3-37. A desk clock with a mirrored acrylic surround. The case is a 6-inch cylinder, similar to the one used on clock with a prismatic center. The case and base are dipped in paint for a marbleized effect. Owner: Nancy Miller.

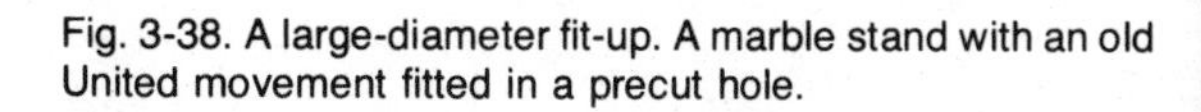

Fig. 3-38. A large-diameter fit-up. A marble stand with an old United movement fitted in a precut hole.

Fig. 3-39. Back of marble clock with a United movement. The back plate and inverted cup are fabricated from acrylic.

of the cardboard is for the tongue that holds the face upright in the slot provided in the base. This should measure ⅜ of an inch deep by 3 inches wide.

Straight lines should now be drawn from point to point where the radiating lines intersect the circle. The pieces outside of these lines are then cut off. Use this template for cutting out the face. A coping saw can be used for this step, but a little area should be left outside the line to allow for filing and sanding. After the faceplate is sanded carefully, draw the radiating lines on it lightly with a soft pencil. These lines will aid in fitting the pieces of quarter-round and the pieces of leather.

The clock mechanism used for this project is battery operated. The shaft that holds the hands is threaded so that it can be put through the faceplate and affixed with a fancy nut. The threaded portion of the shaft should be measured carefully and a hole, the same size, is drilled through the center of the face plate. Use Elmer's Carpenter's Wood Glue to adhere the pieces of quarter-round to the face making them flush with the outside edge all around. Fill in any little gaps that appear between them with Elmer's Wood Filler. This is a good time to fill in

Fig. 3-40. A clock with sand-painted face. The brass dial on a background of colored aquarium sand was adhered to the board with Elmer's glue.

any other gaps and depressions on all of the pieces.

Coat the tongue at the base of the face plate with the wood glue and the inside of the slot in the top piece of the base as well. Glue the pieces together and allow to dry. When all of the filing and sanding is completed, the face (inside of the quarter-round) should be masked completely.

A nice, smooth, quick-drying finish is the next step. Two light coats of Krylon wood primer is suggested prior to applying the finish coat. Allow the primer to dry thoroughly and then spray on two light coats of Krylon interior/exterior enamel. Allow to dry overnight because the next step involves a bit of handling.

Using the cardboard template, cut out six pieces of leather all the same thickness. Three contrasting colors are suggested. This not only enhances the appearance of the clock, but it makes

Fig. 3-41. A hexagonal wooden case with a leather face.

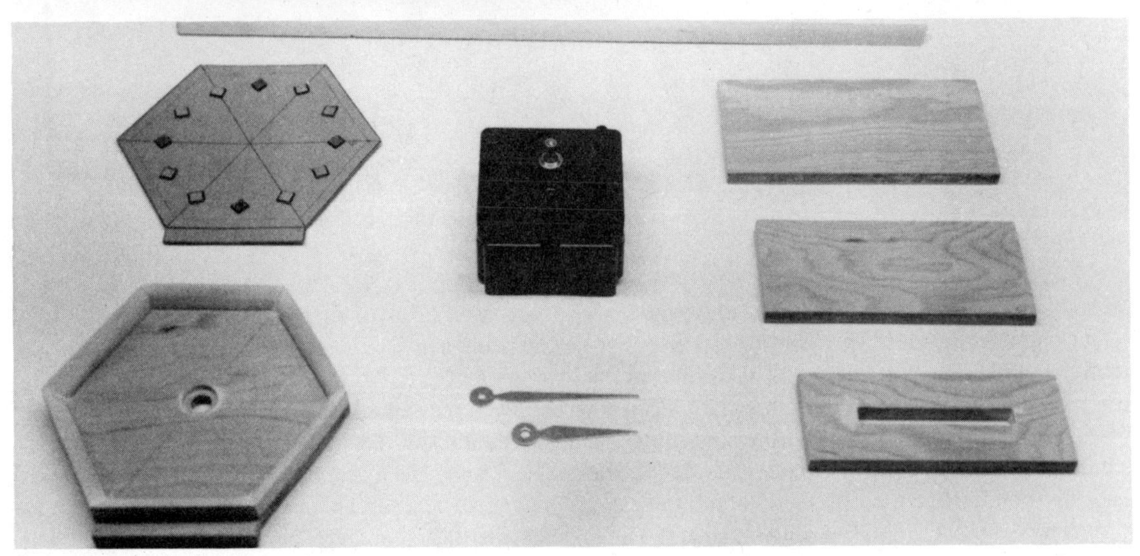

Fig. 3-42. Components for a simple wooden desk clock.

Fig. 3-43. An assembled wooden stand receives several coats of Krylon spray primer.

Fig. 3-44. After the finish coat is added, leather pieces are cut to fit.

the time easier to read. Starting at the top where the 12 will be applied, number each piece consecutively from one through six on the back with a china marker or a bit of chalk. Indicate, also on the back, which point will be positioned toward the center of the clock.

Remove the masking from the clock face. Care should be exercised here so that the painted surface on the molding is not damaged. Coat the whole surface of unpainted wood, from the quarter-round inward, with the Elmer's Contact Cement. Coat the backs of the leather pieces with this cement also, keeping them in numerical sequence. Allow the cement to set for 15 minutes and then carefully apply the pieces of leather, starting with one at the top of the clock face. Each piece of leather must fit within the radial penciled lines when it is applied. Butt each piece up close to the previous one and trim any excess from the other edge with an X-Acto knife. Press each piece down firmly before applying the next one. After the sixth piece has been applied, cut off the points in the center flush with the hole drilled for the shaft.

The numerals and dos that indicate the hours are positioned 1½ inches away from the center of the face. Adhesive-backed vinyl numerals, in a variety of sizes and colors, for this purpose can be obtained from most stationery stores. The ½-inch

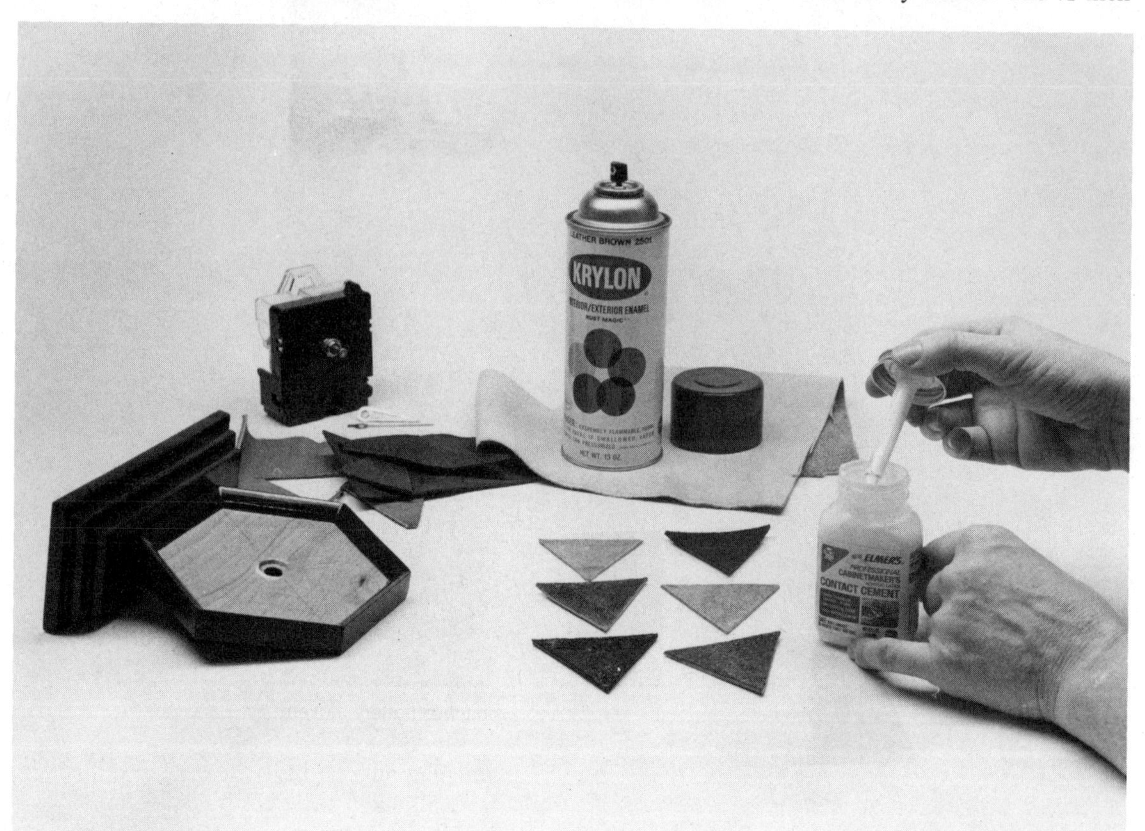

Fig. 3-45. Hexagonal dial and leather pieces are coated with Elmer's Contact Cement in preparation for finishing the dial.

Fig. 3-46. The rear view of a hexagonal wooded clock with a fitted Japanese battery movement.

numerals are best for this particular project.

Now the shaft on the clock mechanism can be inserted through the center hole from the back and the tightening nut can be applied to it. The hour hand is usually press fitted and the minute hand is fastened with a nut on top of it.

Chapter 4

Scratch Building

PRIOR TO ENGAGING IN THE ENDEAVOR OF "DISmembering and reassembling" clock movements, I suggest that you read the sections on repair thoroughly. A good portion of the theory applicable here is spelled out within the pages. The alternating current movement, and the exciting applications to which it can be put, are explored here initially. And there are some slight deviations from the norm in the relationship of pinions to wheels when compared with spring-wound and weight-driven movements. See Figs. 4-1 through 4-5.

In the normal course of wheel-train arrangement, each wheel except the one attached to the spring barrel or the weight pulleys is fitted with a pinion. The wheels drive the pinions on down the line. In the ac movement, power comes from the motor rotor through a shaft to which a pinion is fitted. This pinion turns a wheel whose pinion turns the next wheel. This fact is of vital importance when it comes to rearranging wheels or adding them to produce a visual effect similar to that of a skeleton clock.

Another important difference is that the pinion on the rotor shaft is frequently made to turn the wheel closest to it at just one revolution per minute. The whole movement is simplified in that the motion train is between the plates rather than between the dial and the front plate. It follows that it is a good idea to look for movements with this configuration with which to play. Many General Electric Telechrons are made this way. Some United, Sessions, Spartus, Ingraham, and Westclox movements fall into this category.

Quieter running is achieved when the wheel that the motor powers is made of fiber. These wheels are generally quite visible and lend themselves admirably as pickups for adding eye-catching brass wheels. Many years ago, Seth Thomas used a movement (ac) with a number of fairly large brass wheels resembling those in an alarm clock. The identical movement was to be found in some Westclox specimens. Exposing these wheels made a beautiful timepiece, and especially if the whole alarm mechanism was removed. These movements are no longer manufactured (although Westclox still exists as a subsidiary of General Time Instruments Corp., a division of Talley Industries). Diligent searching still turns up clocks with this movement in flea markets, garage sales, etc. Projects using this mechanism are included in this section.

The inadvisability of changing, for all time, some clocks that are considered "collectibles" right at this moment and fare well to be sought after avidly in the not-too-distant future should be emphasized here. But which ones should you crack open without a moment's hesitation and which ones, when acquired even as bargains, should be set up for immediate display? Here are some simple guidelines.

Anything electric that can positively be identified as having been made before 1930 should not be destroyed. Watching the way avid collectors operate will provide clues. They invariably snap up ac movements that have a little lever or a knob in the rear that must be pushed down or turned to start the clock. These are called *non self-starts* and they might be found with the cases in terrible condition. The ones in mint condition will probably be gone long before you arrive on the scene. Early alarm clocks are now considered collectibles.

Fig. 4-1. A wheel arrangement in an alternating-current movement. The pinion on the shaft from the motor is under stacked wheels and not visible.

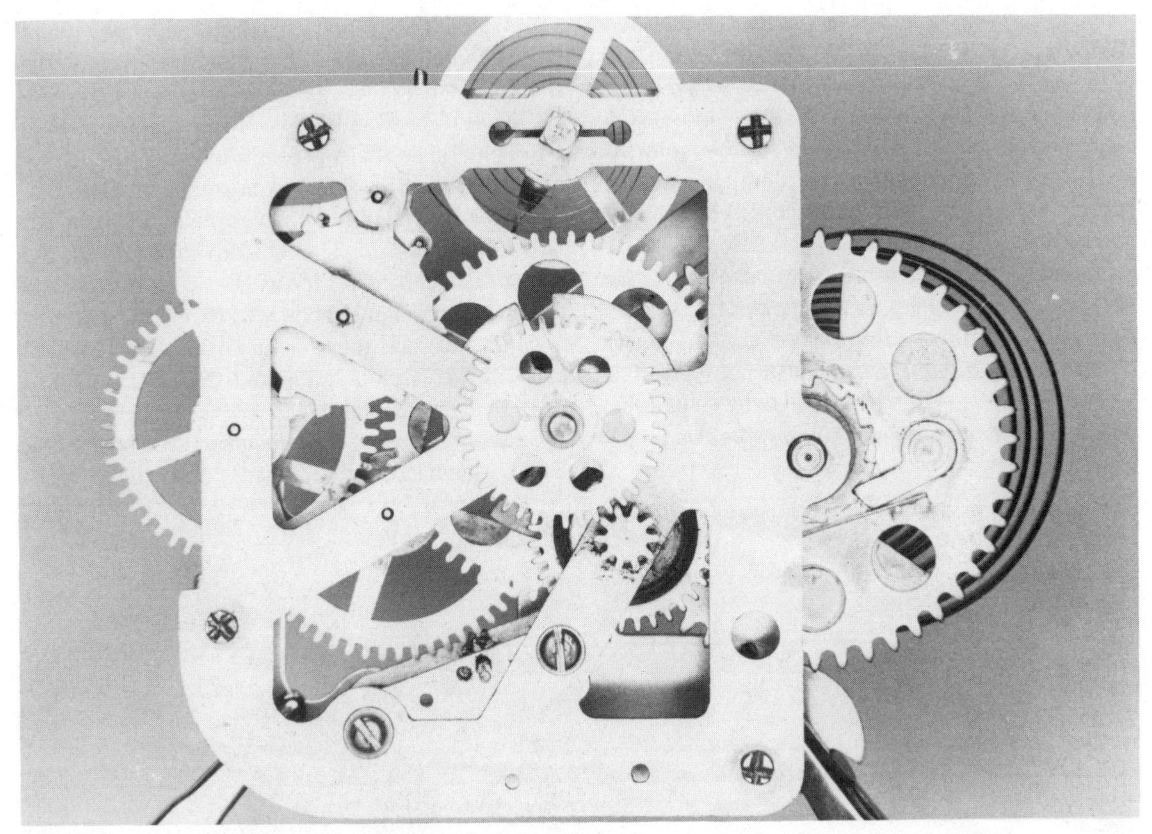

Fig. 4-2. A wheel arrangement in a spring-wound movement.

Early Telechrons might have some value. The story behind them is that the mechanism was invented by a brilliant chap named Henry Ellis Warren. He was awarded a patent for an electric clock in 1909. He formed a company, named the Warren Telechron Company of Ashland, Massachusetts, that he operated for many years with great success. Pieces with this company name, city, and state are real oldies. The company was eventually sold to General Electric who renamed the clocks, "Telechron by General Electric," initially, and then eventually dropped the Telechron on the dial. The word Telechron still appears on the motor inside the movement.

The Hammond just mentioned is a product of the Hammond Clock Company of Chicago, Illinois, which ceased to function as a business in 1936. Laurens Hammond, noted for his electric organ, ran the company. These clocks are non self-starts and some of them have added value as examples of art deco.

Clocks that were manufactured few in number should not be cannibalized. One in this category would be the Atlas Electric, a non self-start in a Bakelite beehive case. No trace of the name of the company appears on it anywhere. It has a beautiful set of large, open brass wheels that certainly should not be hidden! Another specimen in my collection was modified because the original case had been totally destroyed. The movement, made by O. B. McClintock Co., Minneapolis, Minnesota featured solid brass circular plates and two fiber wheels (the extra one to reverse the direction of the motor, no doubt). Correspondence to the company was returned marked "insufficient address." My conversion project is described in this section.

Still other movements (with case) of a more recent vintage—which are worthy of restoration, perhaps, but not working—are the all-plastic novelties copying the blinking eye timepieces of the 1870s. These are usually a figure, a cat, or a dog with a clock in the belly and bulging eyes that move from left to right.

Beside the Seth Thomas/Westclox movement cited above and pieces that could not possibly be restored, what is fair game for conversion to modern skeleton clocks? Well, there are false pendulum GE Telechrons that present a most attractive appearance when modified. A false pendulum is one that does not regulate or power the movement, but is hung on "just for show." Movements with the sweep second hand positioned between the dial and the hour and minute hands lend themselves first to novelty dials and then, with work, attractive displays of the motion and going train mechanism in front of the rear plate.

There is a movement manufactured by the Lux Clock Manufacturing Company, Waterbury, Con-

Fig. 4-3. The pinion on the shaft from the motor shows on this beautiful ac movement with its "spoked" wheels (see arrow).

Fig. 4-4. Collection of clocks picked up at a flea market includes a Hammond (arrow) and an IBM master clock (bottom center).

necticut (now a part of the Robertshaw-Fulton Controls Company) that has *two* pinion takeoffs from the motor and just begs to be treated creatively. Incidentally, the parent company makes a compact, false pendulum movement worthy of consideration. It makes an audible tick like a spring- or weight-driven movement.

Much good work can be done with the early D-cell, battery-powered movements that are large and cumbersome by today's standards. Unusual things can be done with the key-wind, alarm-clock movements.

Much of this work is possible due to the strength and versatility of acrylic plastic in sheet, rod, block, and other forms. Replacing the front or back plates with plastic *does* work whether the idea is pooh-poohed by the purists or not. Clocks with this modification have been running continually for more than five years in our studio. Periodic cleaning, however, is mandatory because you can see the dirt as it collects!

The initial projects that follow are relatively simple and can be accomplished with satisfactory results by even the most inexperienced craftsman.

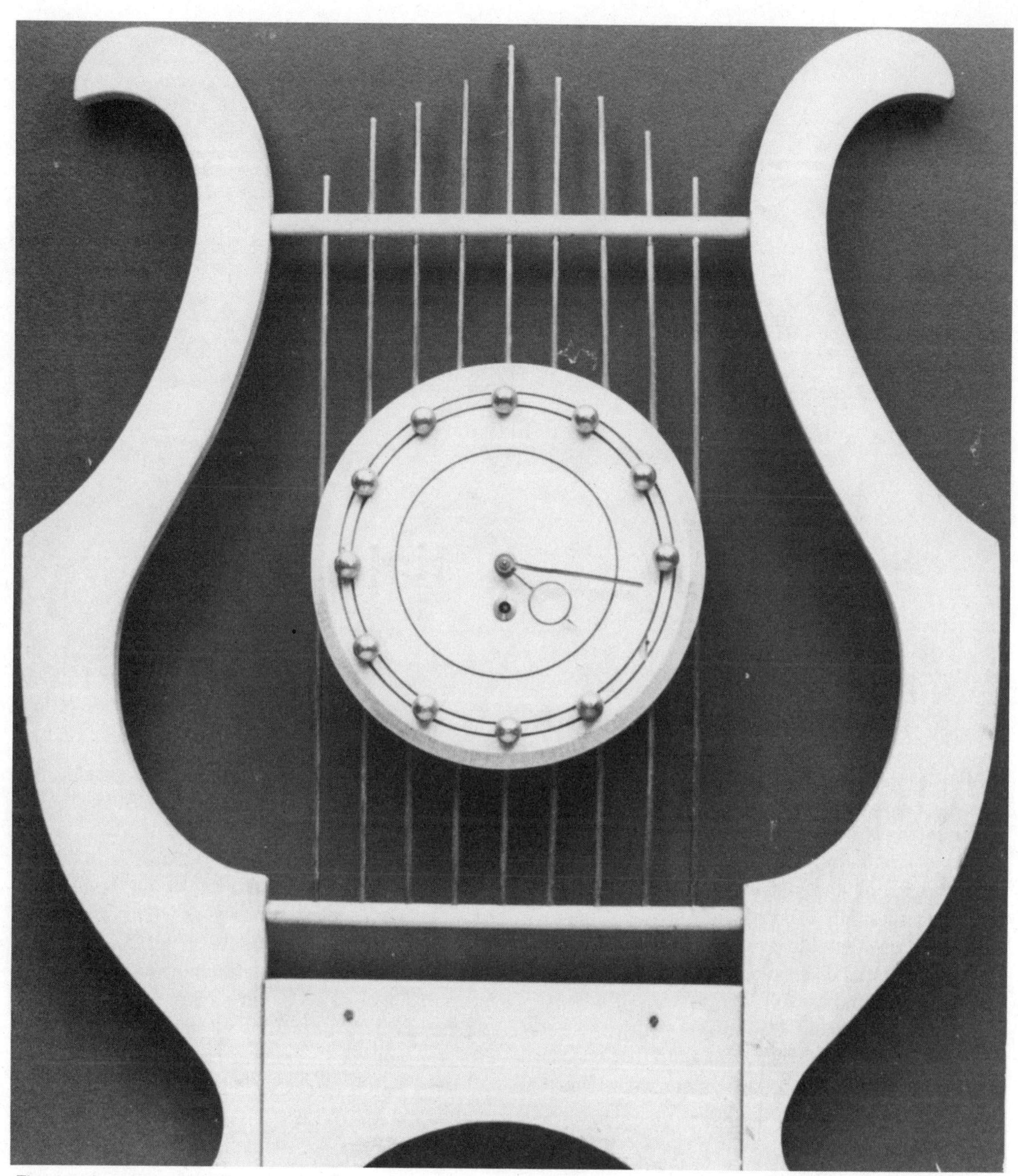

Fig. 4-5. A spring-wound French lyre clock of little value except the brass rods, hands, and the movement for parts.

Chapter 5

Original Dials

IT'S AN OPTICAL ILLUSION! SOMEONE EXclaimed when viewing a reworked revolving-dial Westclox. The sweep second hand is positioned right up close to the dial; it is a feature required to produce this unusual effect. Viewed straight on, whorls appear to flow outward from the center of the disc much as time-lapse photography would show a flower opening. This motion, behind the white hands, is constant and is visible day or night. Simply stated, one large disc 5 inches in diameter—with black lines radiating from the center—revolves in front of another of the same diameter. The inner disc serves as the true dial and does not move. The illusion is created when the inner disc is viewed through the moving one and is heightened immensely if the lines are curved slightly from the center to the outside circumference of both discs. A drawing of one disc is provided. See Figs. 5-1, 5-2, and 5-3.

The stationary disc should be drawn on white paper with the lines curving to the left. The outer disc is cut from thin, clear, rigid plastic and the lines on it should curve to the right. The direction of the lines on one disc must be opposed to those on the other to produce the illusion of an outward flow. Straight lines will produce a pleasing effect though not quite the same.

The installation of the movement was simple because this Westclox had been fitted with a threaded center-mount shaft. A plate of 1/8-inch clear acrylic measuring 6 × 9 inches was prepared with three edges sanded to a matte finish and the bottom edge done to a highly polished mirror finish. Sanding with three grades of sandpaper, 120, 220, and 320, will produce the matte finish. Polishing with 400-grit sandpaper and then buffing turns out the mirror finish.

The highly polished surface on the bottom of the plate is required because the assembly sits over a thin fluorescent bulb. This light unit, salvaged

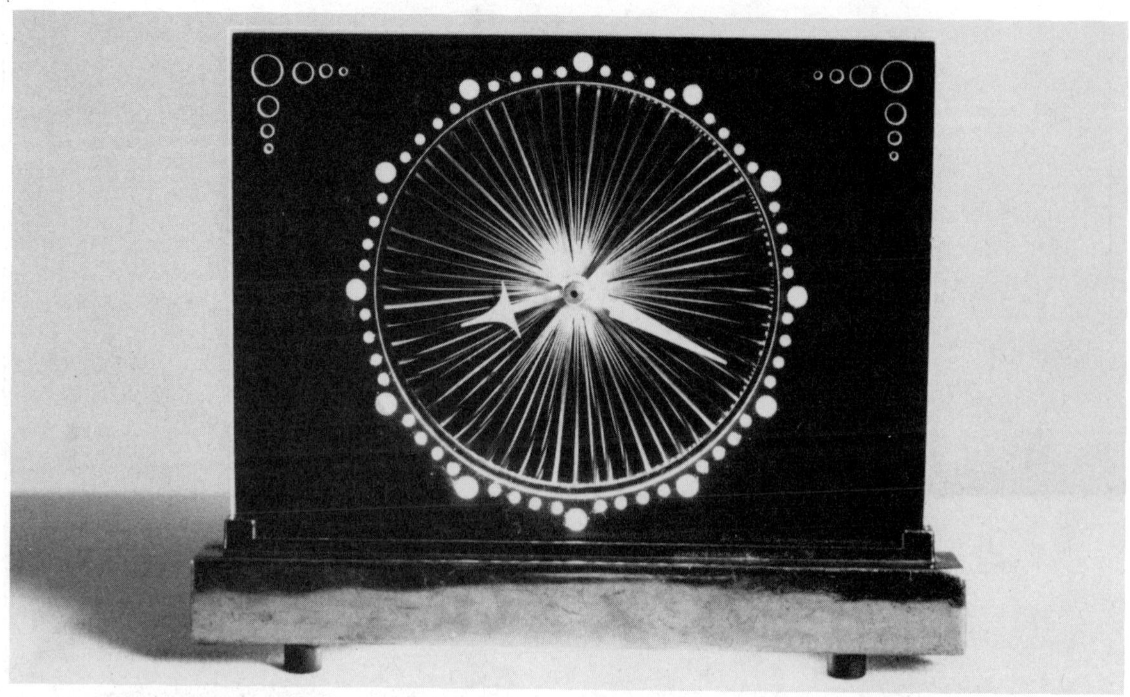

Fig. 5-1. A Westclox with rotating plate fixed to sweep second hand. Unusual visual effect is produced.

from a display sign, illuminates the dots around the dial and the edges of the decorative holes drilled in the upper corners. A piece of ⅛-inch-thick black acrylic, with a 5-inch hole cut in the center, is prepared with polished edges. The center hole allows the movement to protrude to the rear when the two pieces of acrylic are fitted together and clamped into the lamp unit. The whole assembly is held upright in the lamp housing by two knurled screws in the rear.

The outer acrylic disc, with its radiating lines, is carefully centered and glued to the sweep second hand with epoxy cement or cyanoacrylate cement (Krazy Glue or Wonder Bond). The disc/sweep second hand combination is then put on the center arbor, parallel to the inner disc. The hour and minute hands are affixed in the conventional manner, and finally the hand set knob is put on. This movement has a front-hand setting capability.

The ac leads from the movement are wired into the line cord so that the light can be turned off at will without stopping the clock. A wiring diagram for this appears in the fit-up section.

There are all sorts of variations on this theme. One is the cube-on-platform configuration (Fig. 5-4). A disc of prismatic tape affixed to the sweep second hand produces rainbow colors when a light is directed toward it. Try extending the outer disc with its radiating line to a diameter large enough that black dots, matching the white ones, could be printed on it. The white dots would appear to blink like Christmas tree lights. This has been done on watch dials.

ROTATING CHAPTER RING ON DRUM BASE

The concept shown in Fig. 5-5, though unusual, is

Fig. 5-2. Pattern for rotating plate for Westclox.

Fig. 5-3. Swirl pattern for Westclox rotating plate.

Fig. 5-4. Cube-on-platform variation on Westclox rotating disc.

not new by any stretch of the imagination. Ornate drum table clocks quite similar to this were made as early as 1581. One pertinent difference was they were fitted with one hand in the center of the top that pointed out the time on a dial immediately beneath it. In the mid-1930s, the Lux Manufacturing Company of Waterbury, Connecticut issued an inexpensive version of the one shown in Fig. 5-5, but with the circumferential dial duplicating a tape measure.

About this time, pieces resembling this one were issued as promotional items. An acquaintance recalls that, when he was very young, his father brought home one of these promotional clocks from which the acquaintance and his siblings learned to tell time. He remembers distinctly his surprise and temporary confusion upon starting school and being faced with round-dialed, two-handed clocks hanging on a wall that everyone but he could read!

An amber-powder box was inverted to make the base or drum for this one. The original top was reduced in size to fit just above the drum. The concentric circles were etched on the top by

Fig. 5-5. Rotating chapter ring on drum base. The ac movement is positioned horizontally in the base.

Fig. 5-6. This double-decker has an engraved plate beneath the dial face.

chucking it into a drill press and turning it at slow speed with coarse sandpaper resting on it.

A small GE Telechron movement was inserted from below. The minute wheel and collar were discarded and the disc was cemented to the hour shaft. A domed metal cap with a dot of adhesive-backed vinyl stuck on it was positioned on the very end of the sweep second pivot and cemented there. This indicated whether or not the clock is running. The numerals and the half-hour indicators are vinyl. The fleur-de-lis indicator on the base was cut from gold-colored paper and glued to the surface.

DOUBLE-DECKER

In the early 1800s, Elias Ingraham developed a case for clocks that captured the public's fancy immediately. It was referred to as a double-decker because it had a door in front of the dial and another in front of the pendulum. There were versions also with three doors aptly named triple-deckers. The purchaser had the option of a painted lower door or one with a mirror. A diversity of case trim and movements also could be obtained.

This movement shown in Figs. 5-6 and 5-7, with a frame separating it into two parts, came from

Fig. 5-7. The rear of a double-decker clock showing an old synchron movement.

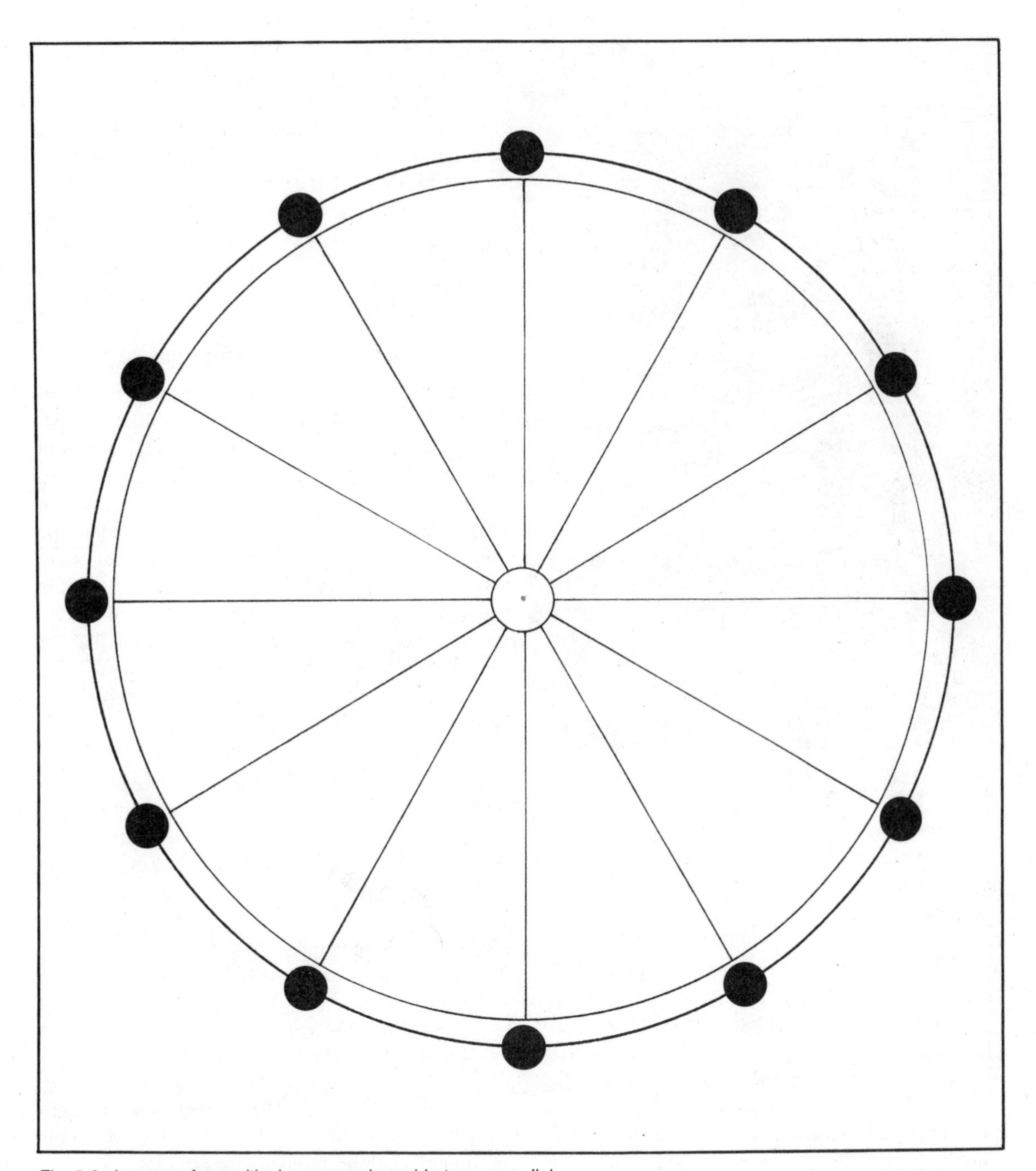

Fig. 5-8. A pattern for positioning numerals and batons on a dial.

Fig. 5-9. An engraved dial in a traditional case made totally of acrylic. Owner: Lillian Smith.

an advertising clock. In retrospect, it might well have been left intact in the hope that its value would appreciate. The movement, however, is not so rare. Trade marked Synchron, they turn up in timers and master clocks (the ones that used to trigger "slave" clocks spread out through a large building).

Seeking to take advantage of acrylic's ability to pipe light, the original piece of glass was replaced with a 4- × -7-inch piece of Plexiglas. The engraving was done on the lower portion with a thin drill bit in a Dremel Moto-flex tool. The black acrylic base has a miniature bulb in a plastic housing that throws all its light up to the bottom edge of the Plexiglas. Figure 5-7 shows just how much metal went into the manufacture of consumer goods in the mid-1930s.

This clock can double admirably as a light for watching television because it is compact enough to be positioned on top of the set or on a windowsill close by. Note the holes in the rear of the light box that allow the heat from the light bulb to dissipate. Acrylics will deform if they are allowed to get too hot.

ENGRAVED DIAL IN TRADITIONAL CASE

This piece can only be considered a labor of love! The engraving took almost as long as the construction of the case. The cross-hatching in the center of the sunburst is applied with a drill bit ground to a point in a Dremel Moto-Tool. The initial and the petals on the flower were done with an impact type engraving tool. The hour indicating dots or markers are the result of pressing a standard metal drill bit in the surface from the rear of the plate at moderate speed. A hole was drilled clear through the plate at the 12 position and a tiny bulb epoxied into the hole. Many movements come fitted with these little bulbs for dial illumination at night.

The clear plate was backed up by one of blue acrylic, $\frac{1}{8}$-inch thick. Mounting the movement involves a technique dealt with at length in the following section on the capabilities and uses of acrylics and other plastics in case making. Briefly, then, a hole was drilled in the center of the flower just large enough to admit the motion train arbor. There must be enough clearance to allow the collar to which the hour hand is fitted to move freely. See Figs. 5-8 through 5-11.

The movement was positioned on the plate and the places where the mounting studs were to go were marked. The original studs, it should be noted, frequently have to be ground off the mounting plate because they are set up for affixing a plate or cup from behind the movement. The holes that

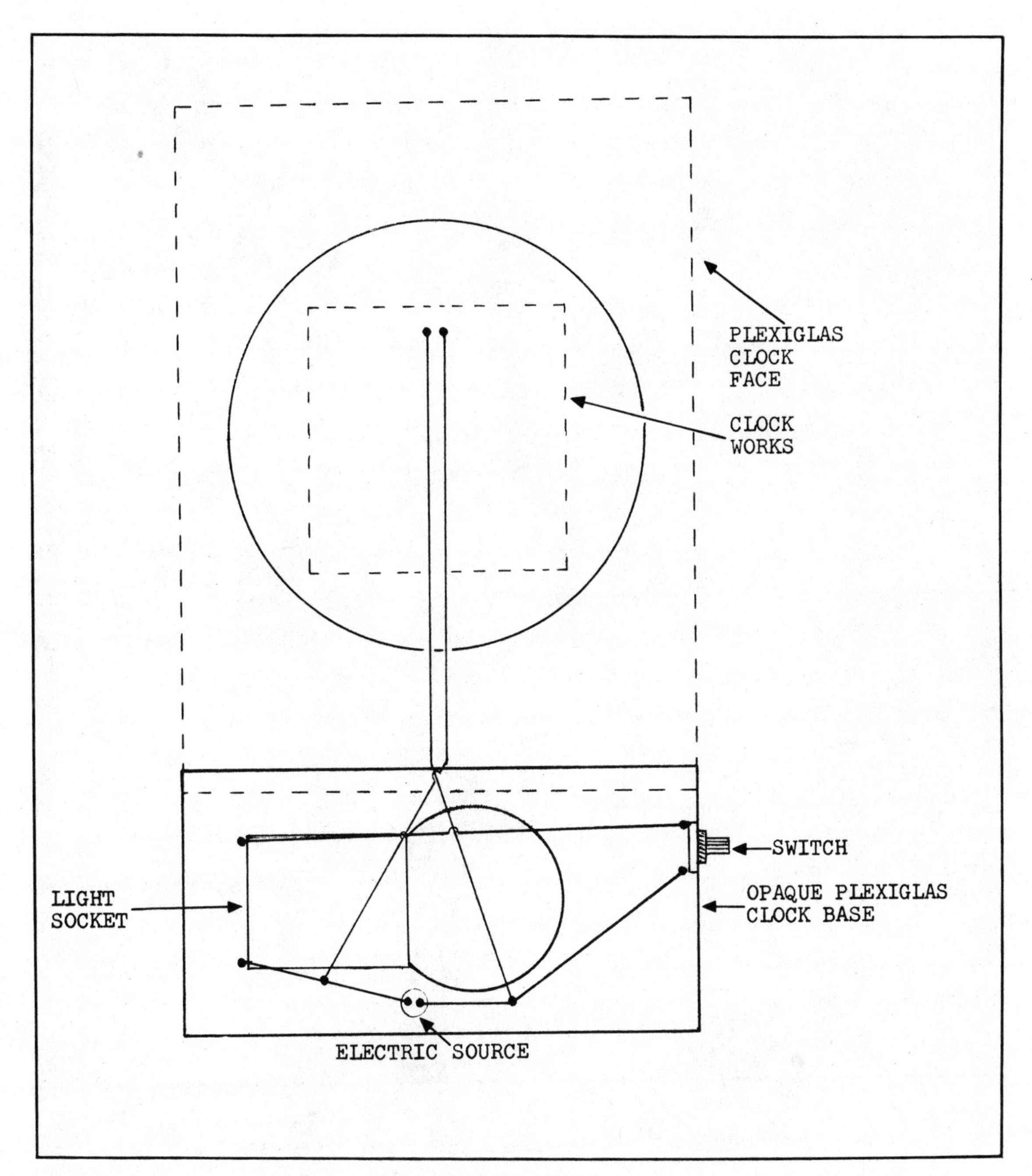

Fig. 5-10. A wiring diagram for inclusion of light in a clock base.

Fig. 5-11. A display of clock faces salvaged when movements were used in other constructions.

are left when these studs come off the plate are the ones used as guides in marking the acrylic plate. The top one is below the 12 position in one of the small petals a bit to the right of the hour hand. The lower one shows on the left edge of the large petal directly above the 6 marker.

While the original pillars between the plates of some movements can be reused often, the use of 6-32 machine screw has proven most effective over the years (and were used in this case). Holes were drilled into the plate with a 7/64-inch drill bit. These holes must not exit the front of the plate. The holes are then threaded with the correct tap. I use one marked NC HS1. The heads of the screws are cut off, the ends are rounded to accept the proper nut, and then the screws are run into the plate. The movement is positioned on the plate and the nuts are tightened on the screws. The plate is then, in this instance, cemented into the case.

The case for this clock is made completely of clear Plexiglas. The pieces were cut to size and the ends beveled to a 45-degree angle. A special high-speed cutter in a router was run along the leading edge of each piece. These four pieces were then cemented together quite like a box with no front or back. Two extra, slightly larger pieces of Plexiglas were given the same treatment, and then cemented to the top and bottom of the case. When this router treatment on the edges is employed, no sanding is required.

Exact dimensions are not given because the size of the case will vary according to the movement selected for use. The overall effect achieved here is that of a case with a traditional look with a crystal appearance in front and complete transparency on the sides, top, and bottom. The front was made opaque in this instance because seeing through the plate would soften the impact of the engraving. The color blue was used at the request of the person who commissioned the clock.

Chapter 6

The Plastics Story

THE ERA WHEN WOOD AND METAL WERE used almost exclusively in the manufacture of clock cases is long past. The Spartus Clock Division of Walter Kidd & Company in Louisville, Mississippi manufactures grandfather clocks with tall cases of polystyrene formulated to resemble the finest of wood grains. The cases look quite as good as wood, company officials believe, and clocks fitted this way sell for a small fraction of the cost of their wooden counterparts.

These officials also tout the scratch-resistance of their product, its imperviousness to humidity (no warping), and the added advantage of no polishing being required. All of this is in reference to Spartus' tall-case clocks, but timepieces in their general line of wall and mantel or table models show a continuing increase in the use of several types of plastic. This includes, wheels, plates, cases, and pendulums.

The Howard Miller Clock Company of Zeeland, Michigan 49464 can be considered a forerunner for the manufacture of acrylic clock cases. Their astute combinations of polished chrome and Plexiglas result in exquisite long-case, table and wall clocks that are ultramodern in feel and appearance though always in extreme good taste. Prices for their products range from $57 to over $2800, and this with the use of wood almost nonexistent!

Spartus is not alone in the large amount of plastic used in their movements. Sunbeam had increased the use of this material over the years and this has been the case with Geneal Electric, Ingraham and, in a very large sense, Westclox.

In addition to clear, transparent plates, the craftsman can produce movement with black, white, red, gray and brown wheels as well as brass and steel. Westclox has worked closely with Du Pont and now uses Hytrel to make the first reduction wheel in all of its electric wall clocks. Hytrel aids materially in the production of reliable, quieter

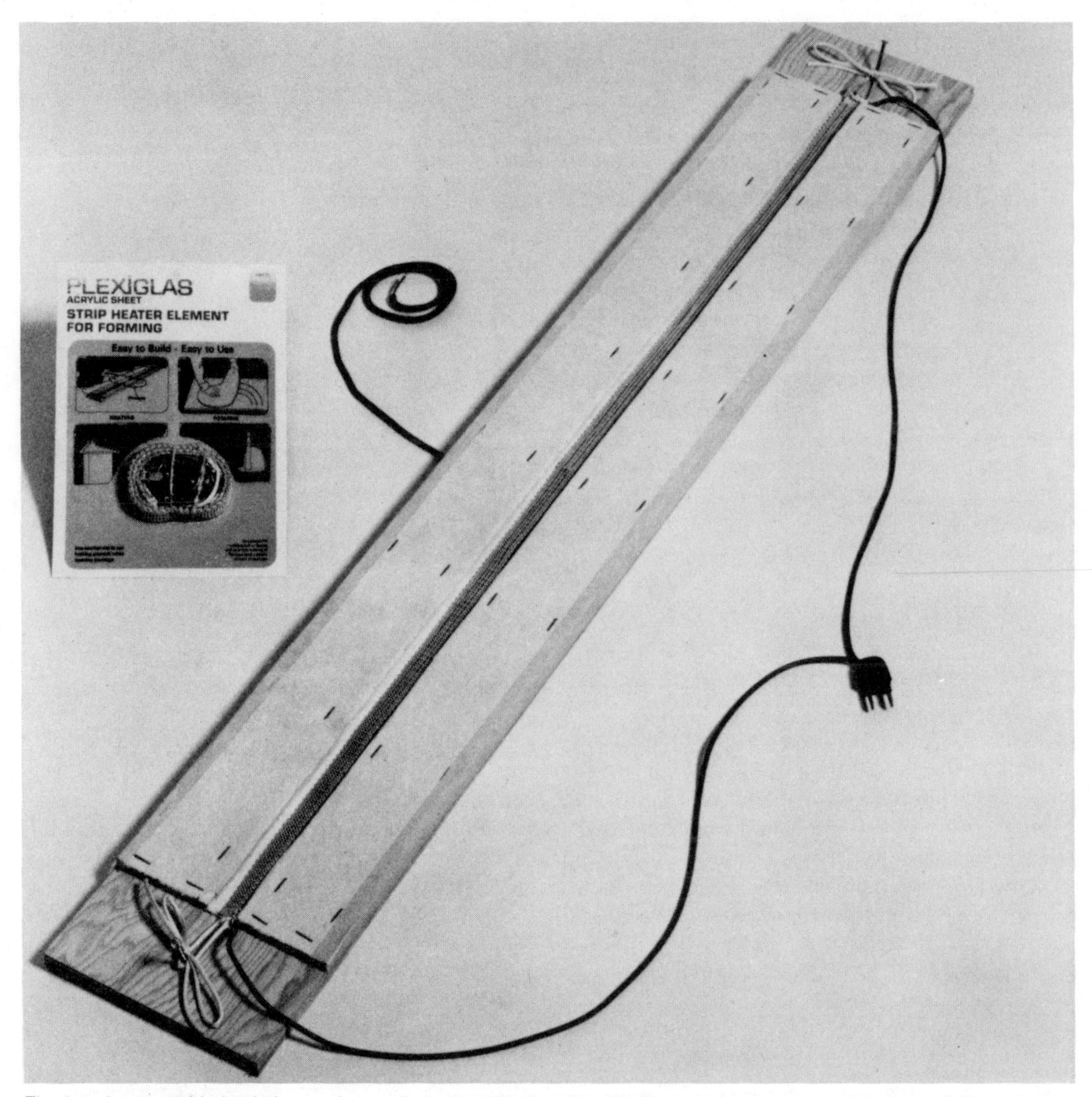

Fig. 6-1. An assembled strip heater for acrylic is made up from the kit shown at left (courtesy of Rohm and Haas Co.).

running clocks with phenomenal longevity. This is possible, the company claims, because of Hytrel's ability to absorb gear chatter generated by the motor pinion.

The use of plastics, as suggested in this book, is of proven feasibility for the craftsman as well as the manufacturer. Hints and tips for the machining of plastics follow. It is not an oversimplification to

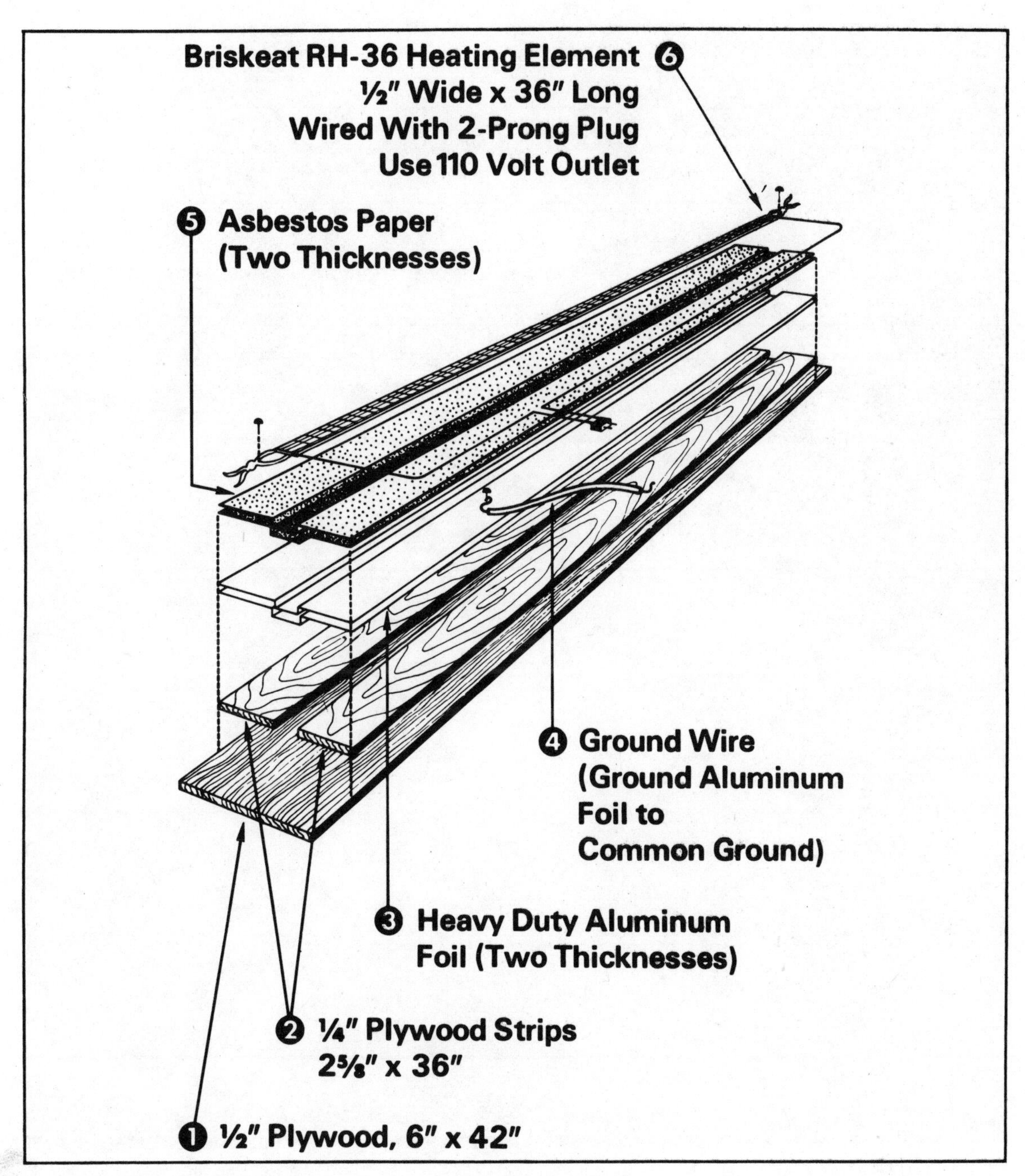

Fig. 6-2. Exploded view of strip heater (courtesy of Rohm and Haas Co.).

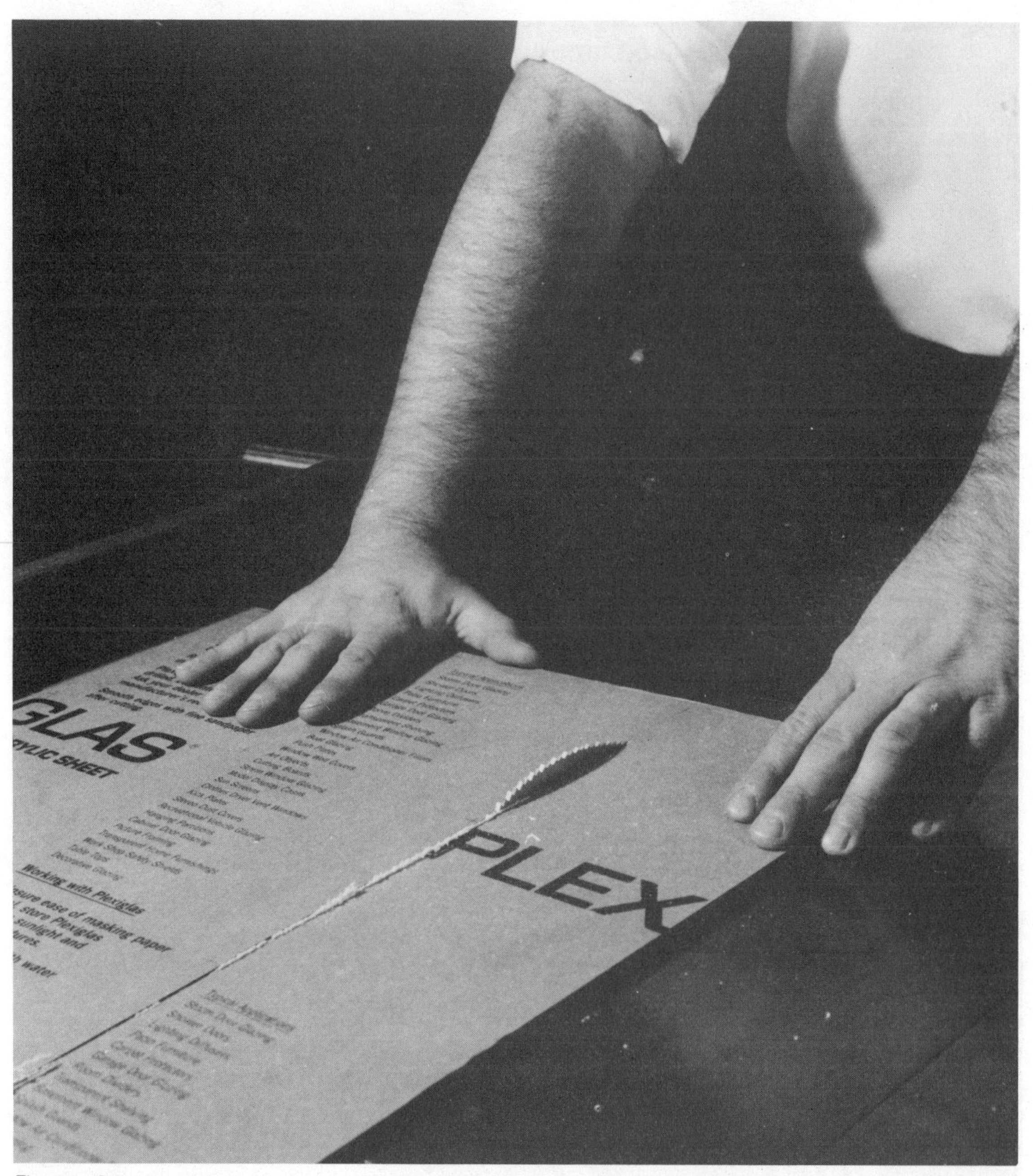

Fig. 6-3. Circular saws are excellent for straight cutting. Cope RH-600 or RH-800 circular saw blade for Plexiglas is recommended (courtesy of Rohm and Haas Co.).

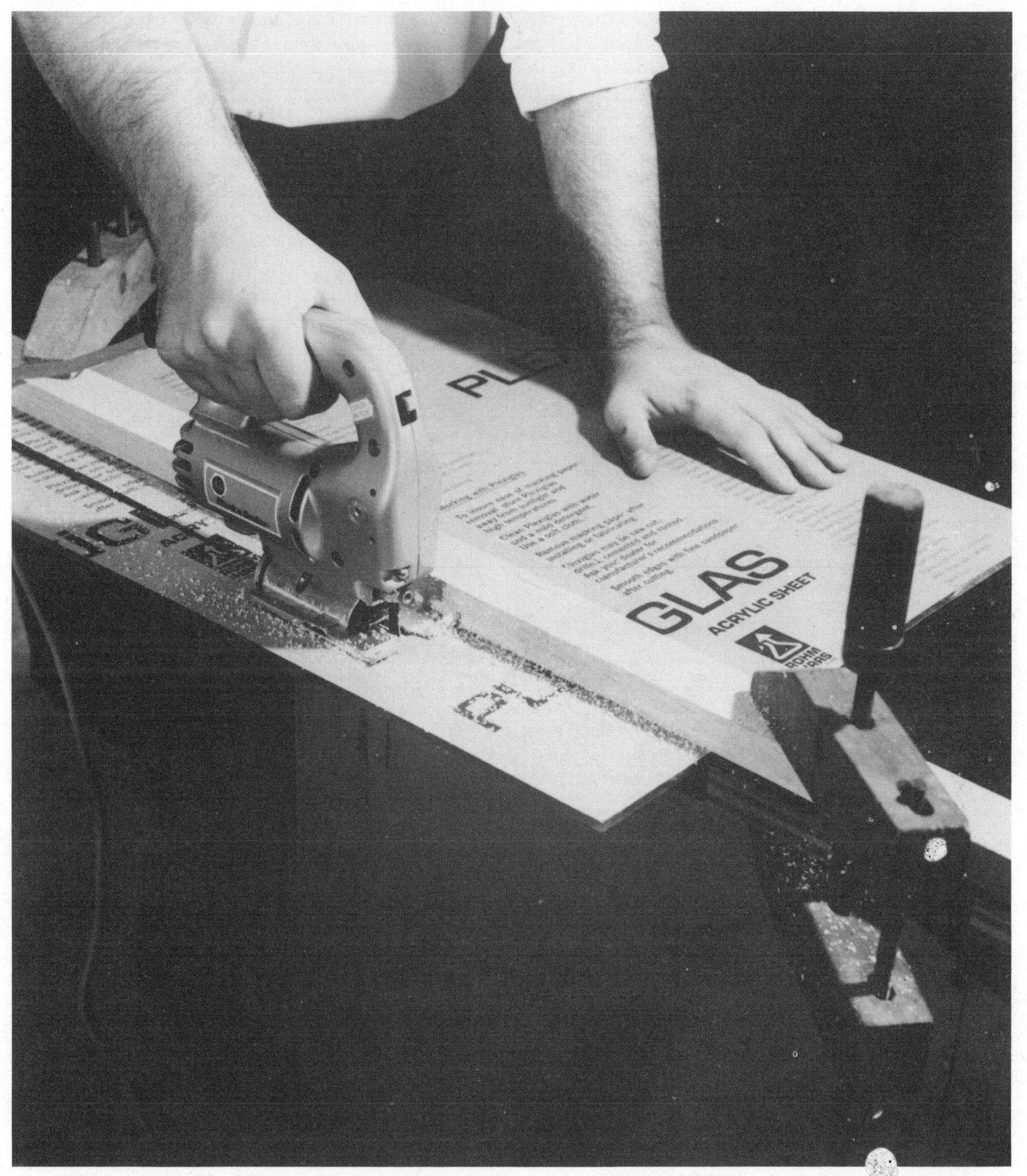

Fig. 6-4. Curved shapes are cut easily with a sabre, band or reciprocating jigsaw (courtesy of Rohm and Haas Co.).

state that plastics can be worked quite like wood and with some of the same tools. It is much more abrasive and that means greater wear on tools. Sawing acrylics requires blades with more teeth per inch and drill bits require a different point to make clean holes without chopping up the material as they exit from the hole. Plastic surfaces tend to scratch easily and require more care in handling for that reason, but they offer an added dimension in that solid pieces of the clear type can be looked into right to their very core! Solid pieces, for example, can be made with color in the center that is visible without any removal of the exterior surface. Compare that with wood, stone, or marble!

ACRYLIC SHEET TECHNIQUES

A majority of the projects included here (Figs. 6-1 through 6-16) were assembled from acrylic sheet cut with either a circular, sabre or jigsaw. It is advisable *not* to take off the masking paper that is normally on the sheet. In fact, lines and drawings can be penciled right on it for cutting. If some sheet

Fig. 6-5. Scribing acrylic sheet, up to 1/4 of an inch thick (courtesy of Rohm and Haas Co.).

Fig. 6-6. After scribing, place a length of dowel under the scored line and break by exerting pressure on both sides of the line (courtesy of Rohm and Haas Co.).

is obtained without the backing paper, the area to be sawed should be covered with masking tape on both sides of the sheet. This is said to reduce friction and to prevent the material from gumming up in back of the blade. The sabre saw or Dremel's motorized scroll saw is excellent for cutting curves. For straight cuts, a circular saw fitted with a Cope RH-600 or RH-800 blade worked wonders.

The best results obtained in my studio when it came to drilling were with Hanson Special Purpose High Speed Twist drills in a drill press. Tiny holes for pinions were made quite frequently with clock drill bits in a Dremel Moto-Tool. This setup was also used for drilling brass and aluminum. For the drilling operation, the acrylic sheet was always backed up with wood. The speed of the drill is of vital importance. With bits up to but not including 3/8 of an inch, 3000 rpm is suggested by the bit manufacturer. Bits 3/8 of an inch and larger require a slower speed in the range of 1000 to 2000 rpm. Extra care should be exercised when drilling at the higher speeds because the plastic sheet has a tendency to climb the drill bit. I received lacerations on all my fingers on the left hand once when a disc I was drilling suddenly rode up the bit.

Smoothing sawed edges is done in a number of ways, I remove saw marks with medium-toothed files, then switch to 120-grit, 220-grit, and 320-grit sandpaper on a block. If more than a matte or satin finish is preferred, I go on to 400-grit sandpaper and then buff with muslin wheel charged or dressed with tripoli buffing compound. This produces the mirror finish so necessary for the bottom of dial plates required to pipe light.

Cementing is any easy operation. Rohm and Haas, from whom I learned many of these techniques some years ago, call it capillary cementing. There are a number of solvents for this purpose, including IPS Weld-On #3 Solvent. I am currently using one called Flexcraft No. 25, obtained from a plastics center in Manhattan, New York. The solvent is applied to satin finished edges with the aid of an applicator (a plastic container with a hypodermic type needle on top of it). The cement, which flows very readily, is applied by running the applicator along a joint from one end to the other. The joint is allowed to dry thoroughly before any further work is done.

There are thickened cements available for joints that are somewhat irregular in fit. A satisfactory substitute can be made up by adding acrylic pellets to the regular solvent, but care must be exercised not to stir in air and form bubbles. This product is applied to one surface, it is applied to the mating surface, and then the joint is held or clamped together until the solvent sets firmly. The joint should be allowed to cure for about two hours before being used. Thickened cement can also be used for filling in scratches. Several applications have to be made with 24-hour drying times in between.

Another method used in cutting acrylic sheet is by scribing and breaking. There are tools (cutters) with which the sheet is scribed along a straighedge repeatedly—as many as seven times for 1/4-inch stock—and then the piece is broken off by bending over a 3/4-inch dowel. This works best when the piece to be cut off is wider than 1 1/2 inches.

There are other cements that will work with acrylic to acrylic and wood to acrylic joints, but these produce nontransparent joints, and none were used for any projects in this book.

SCRAP PAPER CADDY

A 4-inch cube was made up of 1/8-inch acrylic sheet. A 1 1/2-inch segment was cut from the front panel so that the scrap paper could be removed easily. A piece of the same sheet was cut 7 inches long by 3 3/4 inches wide to mount the battery-powered clockworks. The radial lines were engraved on the back of this piece, It was bent on a strip heater at a 15-degree angle about 2 3/4 inches from the engraved end. Then the hole for the fit-up shaft was drilled. The top and two sides sticking up above the box were bevelled and polished.

A small movement, made by the Bulova Watch Company in Germany, was modified to fit the space above the paper receptacle. The original case was removed completely, and that can be done with many of these battery movements. Little 8-inch squares of acrylic were positioned strategically as spacers for the movement so that the balance wheel would not rub against the back of the plate. Modifications such as this will vary with the movement used and the effect desired in the finished clock.

The housing for the C-cell battery was cut from the original case and mounted just below the top edge of the receptacle. It was covered with a piece of mirrored acrylic that effectively hid the battery from view. The leads from the battery box were run up through a little slit to the connecting points on

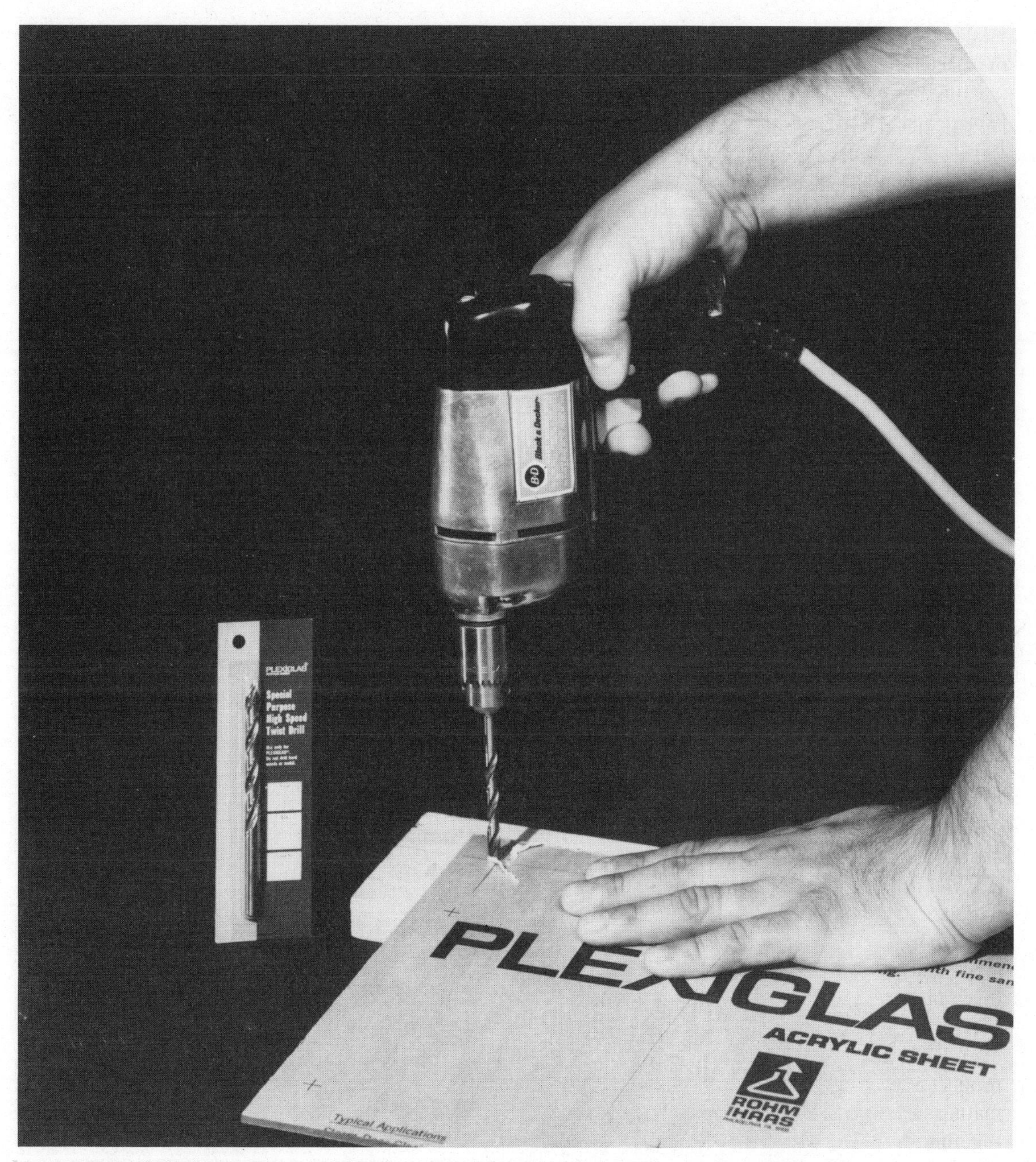

Fig. 6-7. Drilling with an electric drill. Specially ground drill bits (Hanson special-purpose, high-speed twist drills) are required when using power equipment to drill acrylic (courtesy of Rohm and Haas Co.).

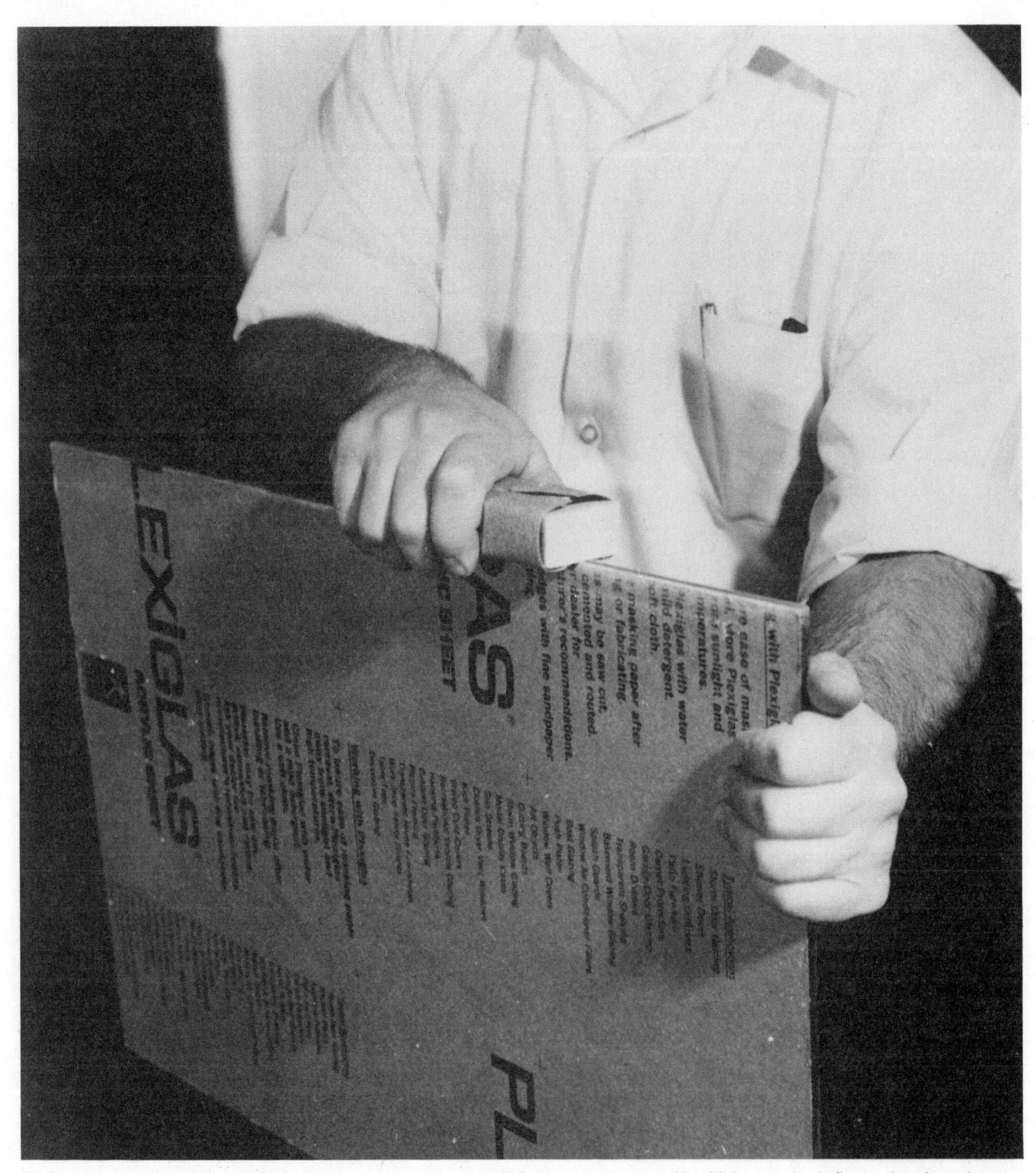

Fig. 6-8. To improve the appearance of the edge and prepare it for cementing, sand it with increasingly finer grits of sandpaper (courtesy of Rohm and Haas Co.).

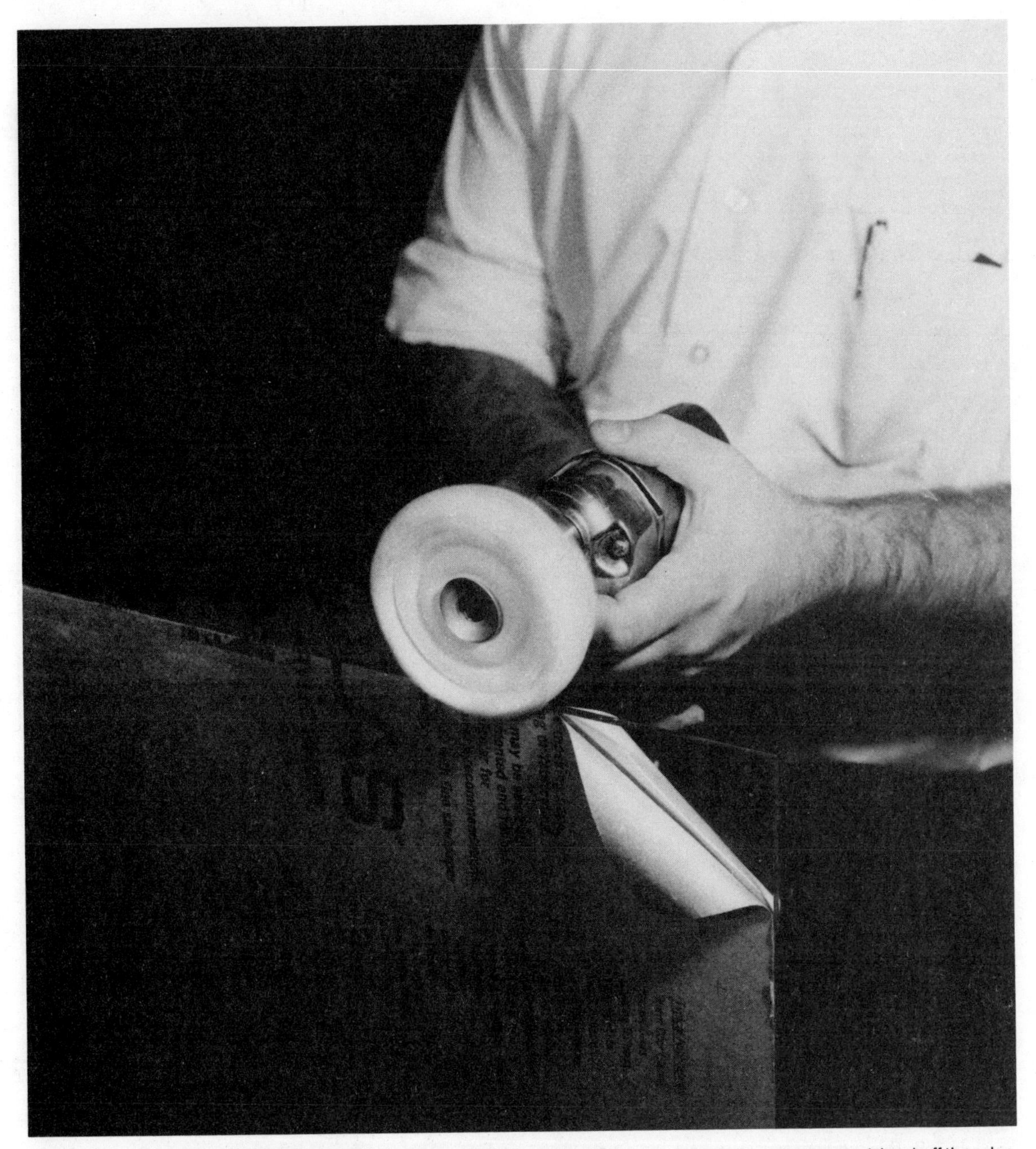

Fig. 6-9. For a transparent high-gloss edge, continue sanding with finer grits (400 to 500) of sandpaper, and then buff the edge with a clean muslin wheel dressed with a good grade of fine grit compound (courtesy Rohm and Haas Co.).

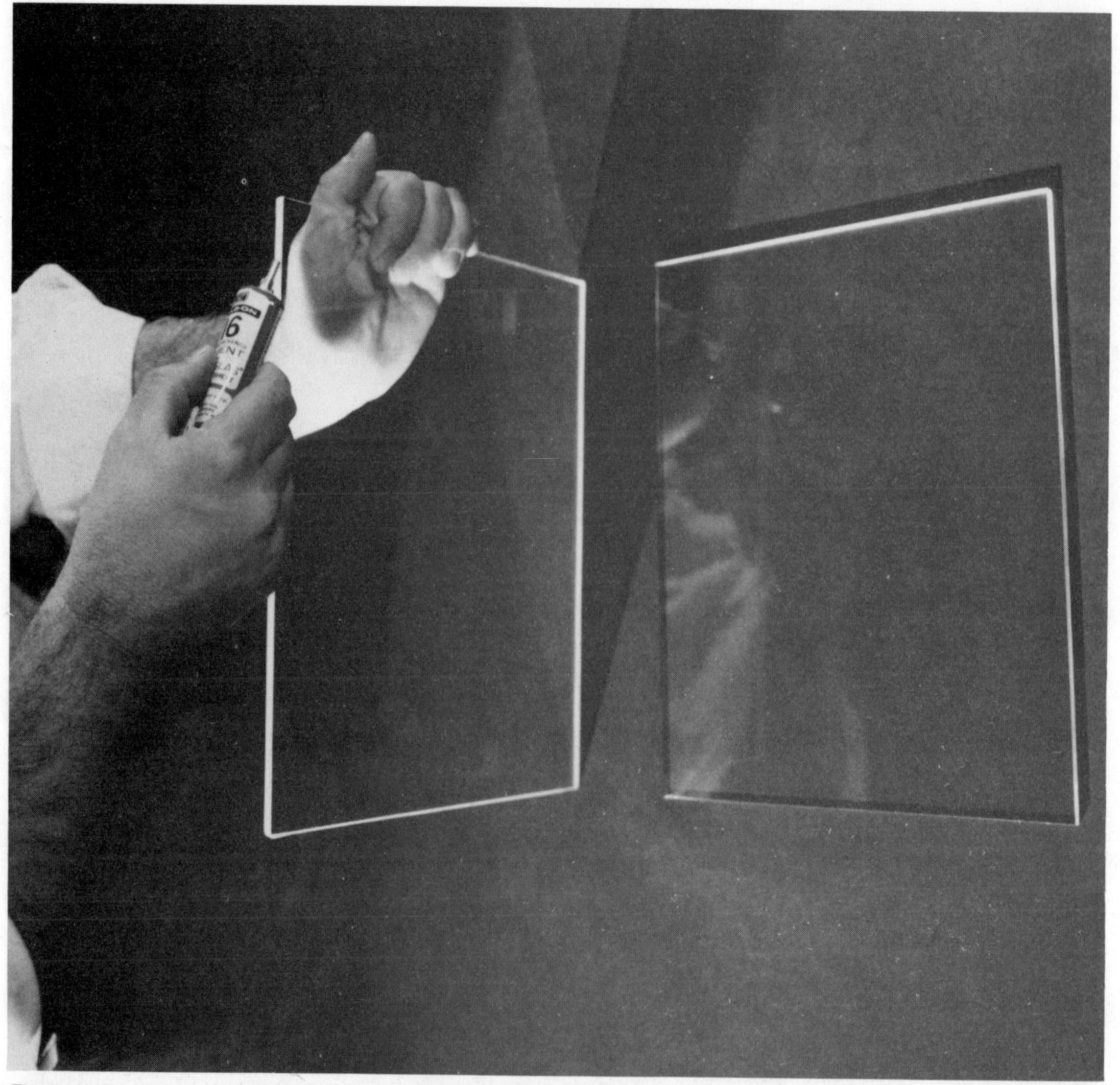

Fig. 6-10. Cementing with thickened cement. Remove the protective masking paper. Apply a small bead of cement to the joint as shown. Join pieces and clamp or hold firmly until set. Let the joint cure thoroughly (about two hours). Courtesy of Rohm and Haas Co.

the movement. One is a natural around to the front plate and the other is soldered to the circuit board just below the transistor.

Many of the older movements of this type will not run when tilted back too far so the deviation from the vertical is crucial and will have to be adjusted by trial with an acrylic strip under the rear portion of the box.

Note that the movement is mounted to the plate with the original screws. Holes just large enough to accept the shafts of the screws are drilled through the plate and then they are drilled out large enough to accept the screw heads. The second drilling goes about two-thirds of the way through the plate. The spacers, fixed to the plate with acrylic cement, actually support the screws. A clear, plastic cover can be made to fit over the movement for dust protection, but it might detract from the overall appearance and was not used in this case. The cover is mandatory, it is felt, when 110-volt ac wires are exposed in such an assembly.

FIBER-OPTICS BASKET

This novelty clock has for its minute dots fiber-optic strands. They surround the movement and lead back from the front plate to a lamp housing some 5 inches away. This distance is predetermined by the flexibility (or lack of it) of the strands.

Fig. 6-11. Painting acrylic. Lacquers, enamels and oil-based paints can be used to decorate acrylic sheet. Spray painting will provide the best, uniform distribution of coating (courtesy of Rohm and Haas Co.).

Fig. 6-12. Scrap paper caddy made up from a 4-inch cube of acrylic sheet.

The strands are available in varying diameters and can be purchased on spools of different lengths. The use of the smaller diameters would produce a more compact unit. The one in this clock was purchased from a plastics supply house, but they can be ordered from Edmund Scientific Co., Barrington, New Jersey 08007.

The movement is a Westclox ac with the front plate cut away to expose the time train as well as the power pinion and the fiber "pickup" wheel. The hands are set from the front and the acrylic dial was cut away in the center to allow adequate clearance for the sweep second hand. The posts connecting the front plate to the lamp housing were cut from ⅜-inch sheet. They are cemented to the front, but drilled and threaded in the rear to accept 4-40 studs cut from machine screws. This arrangement incorporates provision for removal of the back to replace the bulb within the housing. The wiring is described and diagrammed in the fit-up section.

Movement of the fiber wheel is accented by outlining three holes drilled around the center with brass paint. The metal plate and the synchronous motor housing were engine turned to enhance the appearance when viewed from the front.

Mounting to the front plate necessitated substitution of the original pillars with 4-40 machine bolts with the heads removed. The lamp housing is a black acrylic box with ⅜-inch cutouts at the corners to accept the longitudinal posts. A series of ⅜-inch holes are drilled toward the top of the back cover to permit dissipation of the heat from the bulb.

CLEAR-CASED WAG-ON-THE-WALL

This case was designed to display a rather unusual General Electric false pendulum ac movement. GE modified a Telechron unit by the addition of an eccentric plastic wheel propelled by the power pinion from the motor. A pin rides in a track beneath the rim of the wheel and this causes a flat rod to oscillate back and forth. It is linked to a similar rod on the other side of the plate by a similar metal strip from which the pendulum hangs.

All of the wheels and the pendulum assembly were removed so that the plate could be cleaned and engine turned. An extra wheel was added in the upper left-hand portion of the plate to balance the movement of the plastic wheel.

The case, all acrylic, consists of a 4⅜-inch cylinder 2¼ inches deep to which the front plate, 4¾ inches in diameter, and the back cover are fixed. The front plate is cemented on. To facilitate screwing the smoked acrylic back plate on, a ring

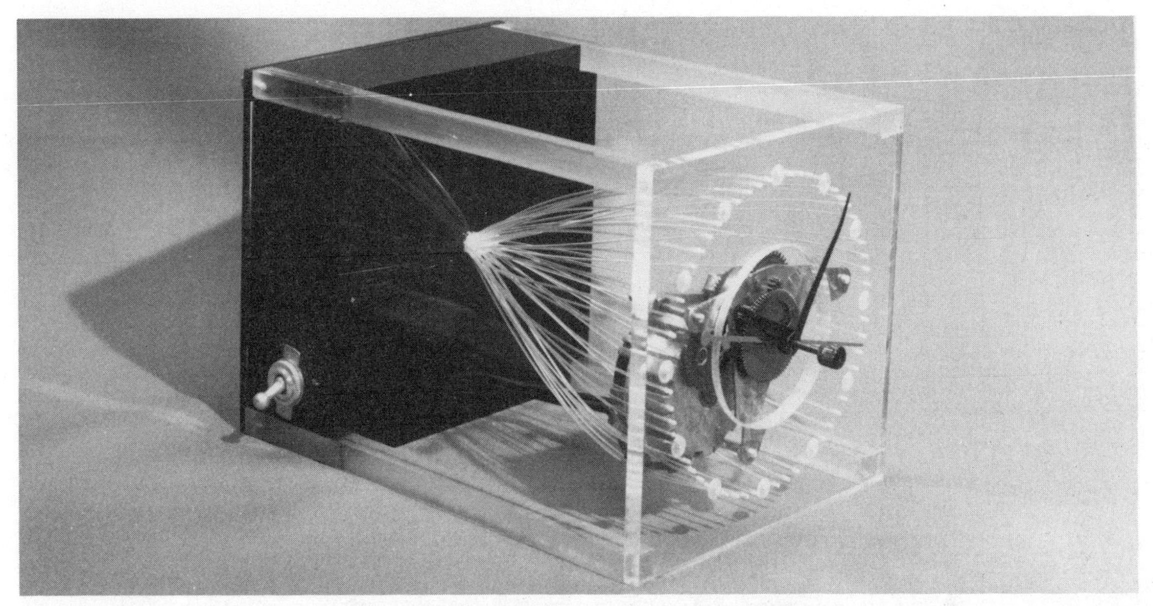

Fig. 6-13. Fiber-optics basket. Novelty with fiber-optic strands lighting the minute dots.

cut from the same stock as the cylinder was cemented inside the back portion of the case. A piece had to be cut out of it so that it could be compressed like an automotive piston ring and sleeved into the cylinder before cementing. The thicker "wall" provided was drilled and threaded to accept short 4-40 machine screws used to retain the back cover. The necessary holes for hanging and for setting the hands were also drilled out of the cover. The finger hole for hand setting is just visible above the six.

Although designed for wall hanging, this piece could certainly be classified as a true "executive pacifier" with all the movement of wheels and pendulum assembly visible. A 1½-inch hole in the bottom of the cylinder permits unhindered movement of the pendulum. Black adhesive-backed vinyl numerals and batons were the final touch.

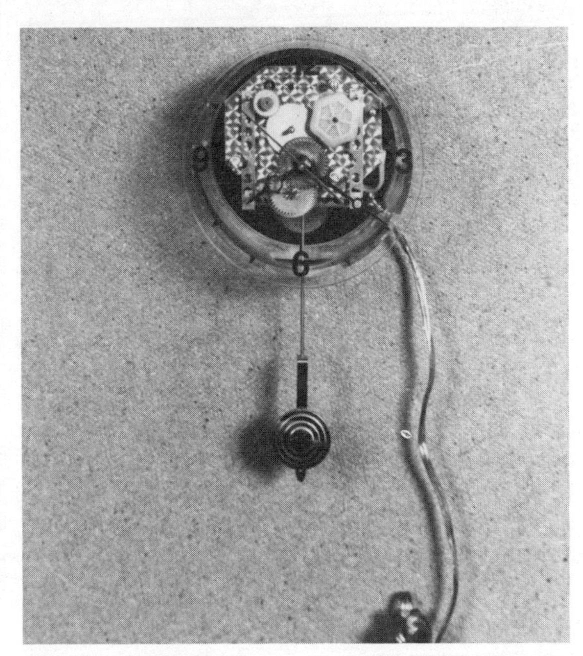

Fig. 6-14. Clear cased wag-on-the-wall, with GE false pendulum ac movement.

BRASS-OUTLINED KEYHOLE CASE

The goal here was to produce a tall mantel clock

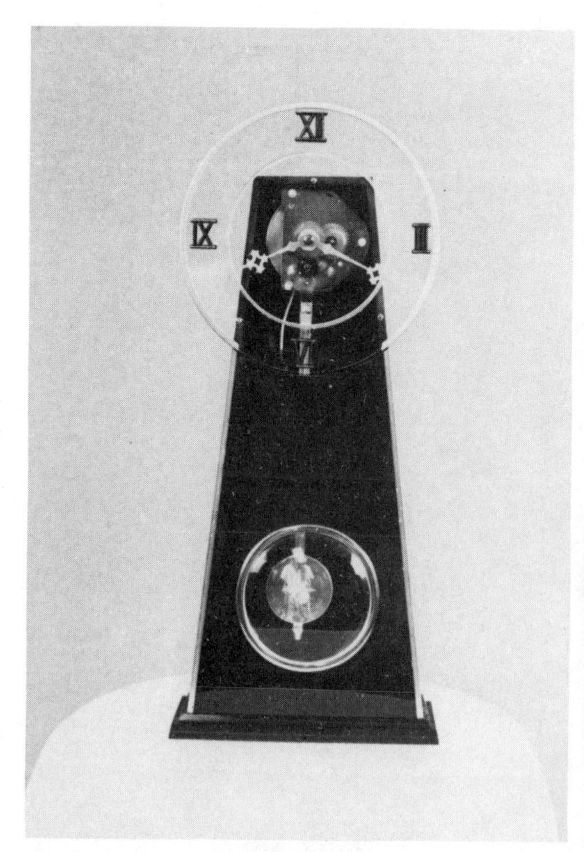

Fig. 6-15. Brass-outlined keyhole case; small banjo-type mantel clock with an ac movement.

reminiscent of the banjo timepieces of the 1820s and 1830s, yet modern in the use of acrylic. The size of the finished case is dependent upon the type of movement selected and, in this case, the length of the pendulum as well as the diameter of the bob. This ac movement, made by the Lux Time Division of the Robertshaw Controls Co., Lebanon, Tennessee came to us with no pendulum or hands.

The Series 2350 motor, with which it is powered, has two power "take-off" pinions. One normally controls the sweep second hand and arbor while the other takes care of the rest of the train. This unit was modified so that one pinion controlled the hands while the other regulated, through a reduction gear, a clever spring-loaded escape wheel.

Many, many of these units must have been produced because exact pendulum and attached anchor turned up in my collection of spare parts. Here, indeed, was a false pendulum that was much more realistic than some of those found in clocks from Spartus and other movement manufacturers. It was used without any additional modification. The metal front plate was removed and used as a guide to accurately drill the required pivot and pillar holes into an acrylic plate.

In the interest of simplicity, just a 4-inch-diameter circle was scored on the back surface of

Fig. 6-16. A swivel dial on a stand. Deceptively simple assembly with the stand bent on a strip heater.

Fig. 6-17. Components of a swivel dial clock showing mounting studs and installation of a neon light at the top of the dial.

the plate. It was drilled and tapped for studs that served as pillars. Aluminum sleeves were fitted over the studs to maintain the proper distance between the plate and the movement. Brass trim, gleaned from a discarded clock case of 1930s vintage, was added to the outer edge of the acrylic disc plate.

Rich, brown acrylic went into the housing. A flat base was bevelled to match the sloping sides. The top of the front piece was cut to conform exactly to the contours of the plate. Lengths of L-shaped edging were cemented to the front corners so that the housing and dial plate were outlined. The movement, attached to the dial with the threaded studs, goes into the housing from the front. It is then bolted on with brass fillister-head screws of the type used on French clocks.

The lower hole in the front of the case (for viewing the pendulum) was fitted with a brass bezel and glass salvaged from the remnants of a small desk clock. The line cord was attached to the leads from the motor and pulled up close to the back wall so that it did not foul the pendulum and the clock was ready to run.

SWIVEL DIAL ON STAND

Simplicity seems to be the keynote with this construction; so much so that viewers have queried as to whether or not it was made from a kit! As shown in the disassembled view (Fig. 6-17) there are but

Fig. 6-18. A swivel case on stand, similar to the swivel dial, but with the engine turning on a black acrylic case.

two parts (dial and stand). The initial step involved cutting the dial plate from ⅜-inch stock with a circle cutter. Material this thick requires a slow turning cutter which must be removed from the workpiece frequently to allow for cooling. The technique of using the front metal plate from the movement as a template to position the pivot and pillar holes applies here. Here, too, all the alarm mechanism has been removed.

Figure 6-18 shows how the back plate was cut away and engine turned. An extra wheel, picking up its energy from the motor pinion, has been added to cover the blank area on the plate. Determining the precise position for the pivot holes for this wheel required the use of good dividers. Not visible when the clock is running is a smaller wheel on the same arbor with teeth matching those on the fiber wheel. This smaller wheel contacts the motor pinion.

There are minor variations to the manner in which each of these clocks are assembled.In this instance, tight fitting plastic sleeves were added to the mounting studs as spacers between the dial and the movement. The little light at the 12 position was epoxied into a cube of acrylic and this, in turn, was cemented to the dial.

There is no case whatsoever for this assembly. When dirt accumulates it is apparent immediately. A thorough cleaning about every 18 months is required even though the clock runs on and on with all the wheels catching the eye as they turn.

An additional thick, plastic wheel is usually employed to set the hands on similar movements, but this was discarded in favor of moving the hands manually. Its replacement would be mandatory if the clock were to be sold or given as a gift. The sweep second hand would be knocked off immediately by someone touching it for the first time.

SWIVEL CASE ON STAND

The similarity between this clock and the swivel dial clock is apparent immediately. One of the differences, however, is that machine screws with knurled heads were used to tighten the case in the stand. In the previous project acorn nuts were used. Engine turning was done on the front of the black acrylic case in this instance. This technique is never as effective on plastic because the small metal brush used tends to "load up" with residue and rapidly at that. The dial plate is cemented to the case and a little light is concealed within, creating a very dramatic appearance at night. The only change to the original movement was the removal of the front metal plate. No wheels were added. It is affixed to the front with two machine screws for which holes in the plate were drilled and threaded. All of the assembly and modification methods employed here are detailed in previous projects.

CUBE ON CUBE CASE

The nucleus of this case is two 3⅞-inch clear photo cubes. See Figs. 6-19 through 6-22. The one on the bottom has squares of black acrylic with bevelled edges (except for the bottom) cemented to the sides. They are slightly smaller so that light from a standard lamp socket with 7 ½-watt bulb emanates from the corners. White acrylic (all ⅛ of an inch thick) makes an inner housing for the movement.

Fig. 6-19. A cube-on-cube case with a brass plate and wheels of an older Ingraham movement exposed.

The white inner box is but 2½ inches deep and open at the back with a piece of 1/16-inch clear acrylic serving as the dial plate.

The movement is one of the older brass plated ac units by Ingraham. The front plate was removed, but the original pillars were retained for mounting to the dial. This movement had one of the time-train wheels fixed to the front plate. A small screw was substituted for the original pin and the pinion that rode in front of wheel threaded to accept the screw. A clear plastic spacer was cemented to the plate in front of this wheel to maintain it in a parallel position to the plate. Locktite was used on the threads of the screw so that it would never back off.

One brass wheel was added in the lower portion of the movement. It has the same number of teeth as the wheel fitted to the sweep second hand arbor that is turned by the motor pinion and makes, therefore, one revolution per minute. Four holes were drilled into this "add-on" so that it does not blend with the brass plate and is readily visible to the eye. Despite temptation, no additional wheels were added.

The movement was deliberately positioned high in the inner housing to allow clearance for the bottom wheel. A little pinion was cemented below the numeral 6 to achieve better overall visual balance. The appearance can be enhanced additionally in this type of mounting by running the brass nuts up on a machine screw, chucking it into a Dremel Moto-Tool or a flex-shaft, and grinding off the corners on a file or coarse sandpaper.

Three-dimensional, adhesive-backed plastic numerals were set on the dial plate and black spade hands were fitted for contrast with the brass plate. The light in the lower cube can be switched off from the back.

POLYESTER RESIN TECHNIQUES

Although there are but one or two projects with polyester resin contained herein, this medium offers the possibility of fabulous cases and dial plates shot through with color. In a dial plate, the color can be centralized so that it just covers the movement in back of it, creating the effect of its "floating" in the plate. It can "wrap around" a whole alarm clock to simultaneously present a feel of color in depth. This effect is achieved in the automobile world by applying 24 coats of lacquer over candy-apple paint. This resin can be molded into unusual shapes and in varying thicknesses whereupon it can be "worked" like wood or acrylic. Thick pieces are cut best with a conventional cut-off wheel, but a hacksaw with 24 teeth per inch on the blade will work admirably.

Polyester resin is most frequently used for encapsulating or embedding objects. No problems should be encountered here provided the proper proportions of resin and catalyst are adhered to religiously. See Figs. 6-23 and 6-24. In making a dial plate, for example, the first step would be to select the proper mold. Liquid casting, as it is called, has become quite popular over the years. Books have been written on it and there are numerous manufacturers who offer glazed ceramic molds in various sizes for this craft. Heat-resistant ovenware, glass dishes, tumblers and other items from the kitchen can be used for molds. Bottles cut to the right height make excellent molds, and even metal candy molds can be added to the list. I used a discarded tin canister in which bulk photographic film had been purchased, fully aware that it would have to be cut and pried away from the molded object once the resin had set permanently.

A mold release agent is poured into the mold to coat it. Then thin layers of the resin are poured. The formula here is two drops of a catalyst (methyl-ethyl-ketone peroxide) to one ounce of the resin. Shallow pourings of the resin are suggested because the mixture generates an exotherm heat that can cause fracturing. Dyes and pigments are added for color; the dyes produce translucence while the pigments provide solid colors.

The preparation of a 6 inch diameter dial plate is as follows. Two ounces of resin with proportionate catalyst added is poured into the bottom of the mold and allowed to gel—not set firmly. When the

Fig. 6-20. The original dial of the Ingraham movement used in the cube-on-cube case.

Fig. 6-21. An Ingraham movement with the dial and dial plate removed. The hand-setting assembly was discarded.

surface of the pour can be touched gently with a toothpick—drawing up a fine thread—embedments can be placed upon it and the next pour can be added. This procedure is continued until the desired thickness is achieved (no more than a half-inch).

The product is allowed to dry for about 40 minutes, whereupon it is removed from the mold. If the sides of the mold are straight up or sloped outward, release can be effected by inverting the whole pour and tapping the rim of the mold gently on a piece of wood or thick glass plate. The top surface of the product (the last pour) can now be sanded and polished. The procedure will vary slightly depending upon the size of the mold, the size and type of embedment, and whether or not the ambient room temperature has been maintained at about 72° Fahrenheit. Lower temperatures will take longer to gel. Higher temperatures can accelerate the production of exotherm heat.

Success in working with polyester resin is assured if these guidelines are followed:

☐ Use plenty of ventilation when pouring resin. Do not work in close quarters.

☐ Use dye and pigments sparingly. Just a drop or two will supply color of satisfying depth.

☐ Use a good grade of resin to avoid yellowing or excess shrinkage.

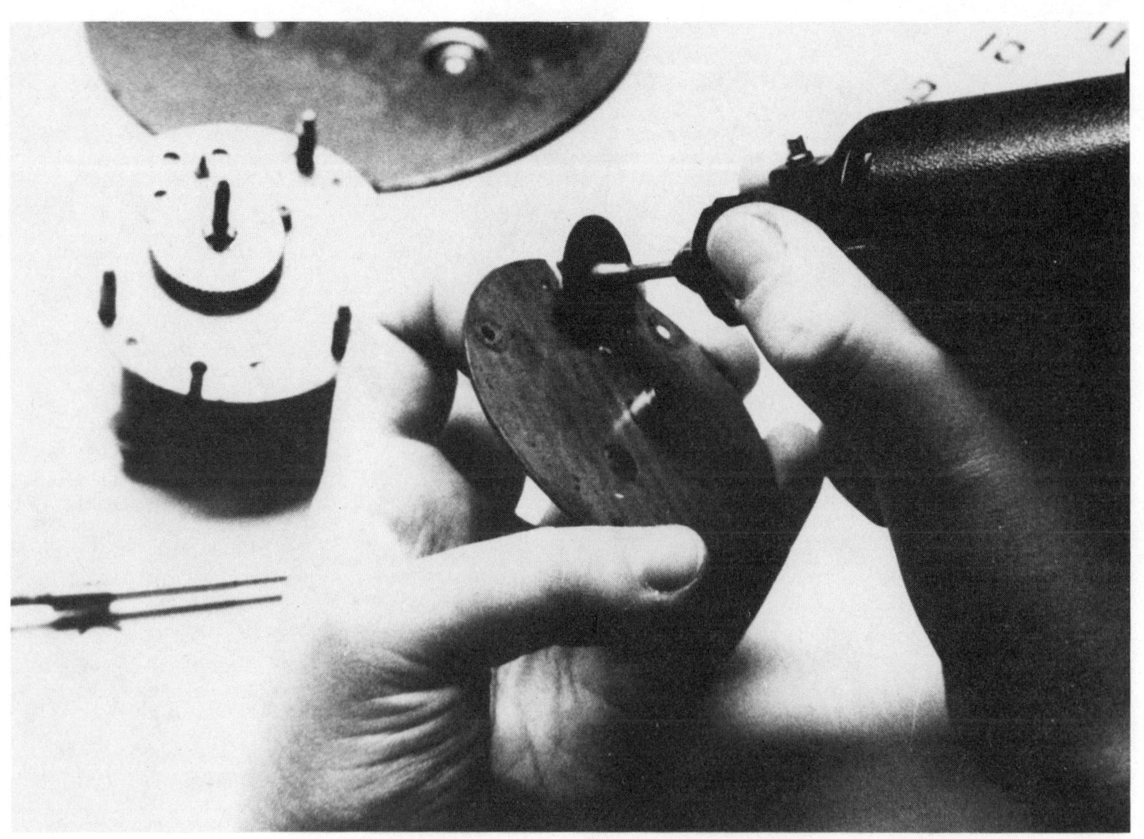

Fig. 6-22. A cutoff wheel is used to free the arbor of a wheel pinned to a front plate. The wheel is fixed to a new acrylic plate with a brass screw.

☐ Maintain even room temperature and avoid dust.

☐ Stir resin gently to avoid excess bubbles.

☐ For additional luster and to aid in dust removal, coat the finished piece with a paste wax (such as Simoniz) and polish with a soft cloth.

SMALL ALARM CLOCK STAND

Start with a bottle or jar about 6 inches in diameter and approximately 5 inches high. Pour about 3 ounces of resin and allow to gel. Carefully center another bottle, about 4 inches in diameter, right on this first pour and make additional pours between

Fig. 6-23. A small alarm clock on a stand. The case and stand are molded separately of polyester resin.

Fig. 6-24. A polyester dial plate. Shot through with color, this plate could be mounted on a stand or backlighted in a case.

the two bottles. Make another pour using clear resin. The outer bottle can be tipped to about a 30-degree angle and propped in this position for the rest of the pours. Add a drop of dye to the next 2-ounce pour and drain this resin into the space between the two bottles. Rotate the bottle and use dye of contrasting color in the next pour. Repeat these steps until no more resin can be added without spillage. Set the bottle upright for one last pour of clear resin and allow to dry thoroughly for an hour or more.

Break the inner bottle with a hammer and carefully remove all the bits and pieces of glass with a parrot-nosed pliers and a tweezers. Put the outer bottle in two plastic bags and break the glass as gently as is possible. Brush off any remaining glass before proceeding to the next step.

The initial, clear pouring that is now the bottom of the resin vessel, is removed out to the walls and filed or ground to accept the alarm clock from which the legs and stand have been removed.

A candy dish was used to mold the base with no embedments, but with matching dye added to each of the repeated pourings. The finished pouring fell right out of the mold when set with no urging. The portion to be affixed to the "case" was sanded flat at an angle. A flat spot to match the base was filed and sanded on the case where the base would be fitted and then glued to it. Epoxy glue can be used as the adhesive or a highly catalyzed ounce of polyester resin can be substituted. Apply with a spatula or a wooden stirring stick to both surfaces. Hold until both pieces adhere completely. Apply epoxy to the rim of the case and fit clock as shown in photograph. Winding and hand setting is done through the wide opening at rear.

POLYESTER DIAL PLATE

The basic procedure for this plate is similar to the one just described. A 2-ounce clear resin pour is used to cover the bottom of the mold. The tin canister (5 inches in diameter) mentioned previously was used here. The random splashes of color, both pigmented and with dye, were achieved as a by-product of a series of ongoing polyester projects. In essence, the mold was used as a draining receptacle for the wooden stirring sticks from a number of simultaneous pouring projects that lasted a day or two. Periodically, an ounce or two of clear resin was poured into the mold to cover the color drippings quite as if they were embedments. When the pour rose to the desired height, one last clear pouring was added and the disc was allowed to set permanently.

The surface obtained at this juncture is never perfectly smooth and, because the resin is exposed to the air, might take hours to cure. Optimum smoothness and accelerated drying time are both achieved by the simple expedient of laying a flat sheet of Mylar, about six-thousandths of an inch thick, smoothly across the surface. The Mylar will separate when the surface is completely dry.

For a plate this size, a compact unit with a ¾-inch shaft is best. Plain bar or spade hands will probably be the most visible. Vinyl numerals will adhere readily to the surface or the digits can be engraved to pick up light, as does acrylic. Note that, in this case, the engraving should be done on the front surface of the plate. Drill the hole for the shaft from front to back where any edge chipping will be far less visible.

A plate, as thick as suggested, will stand alone on a simple, dark base or it can be mounted dramatically by allowing it to project from the front of a case built to compliment it. In the latter event, it can be lighted from behind using methods applied on clocks described previously.

STYROFOAM TECHNIQUES

Styrofoam is a proprietary brand name for expanded polystyrene. The product, as we use it, is the result of expanding tiny beads of polystyrene and fusing them in molds. It is generally available in sheets, cylinders, blocks, cones, and balls. It will not rot or become moldy. It is not irritating to the skin and it does not flake. It *is* extremely inflammable and produces a fine dust when sanded that should never be inhaled. It can be sawed or cut with a knife, razor blade or, optimally, with a hot wire or with a battery-powered cutter.

Fig. 6-25. Fountain of time and wall clocks of Styrofoam (expanded polystyrene).

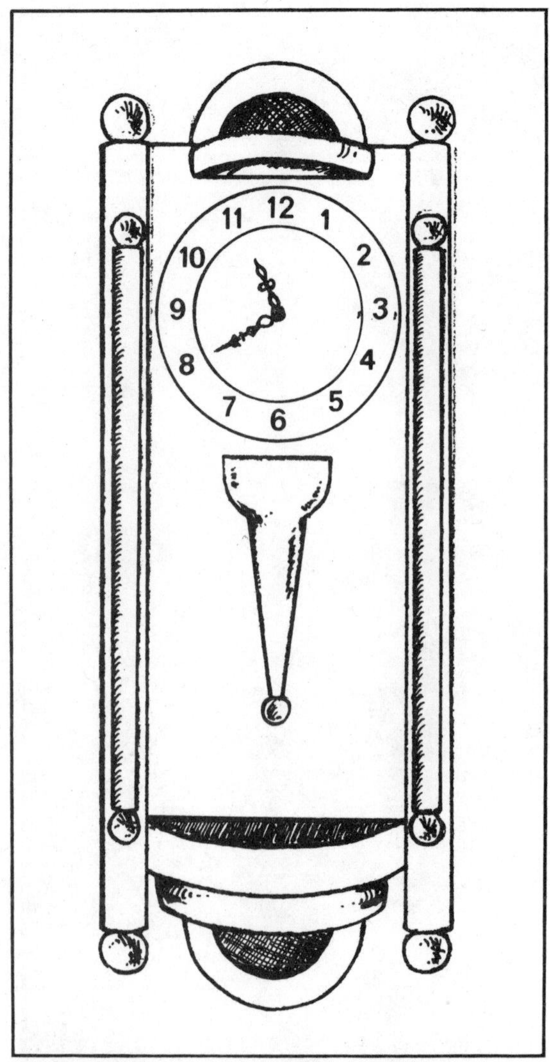

Fig. 6-26. A sketch of a fountain clock showing placement of Styrofoam parts.

The projects shown in Figs. 6-25 through 6-31 were made with the commercially available shapes mentioned before so that the work requiring tools is amazingly simple. One of the most effective ways of sanding this product is to use a hand-held block of the product itself—that is a block of Styrofoam!

Joining pieces is best done with a water-base glue. Elmer's glue is one of the best for this. Large pieces of the material can be joined with toothpicks.

Fountain of Time Clock

Materials Needed:

Cutter (hacksaw blade, serrated knife, or hot-wire cutter)

Nylon bristle paint brush

Wire cutters

Scissors

18-gauge, bare-stem wire

Gesso

Soft rags

Two 1-×-12-×-36-inch sheets

Two 1-×-8-inch-diameter discs

One 2-×-2¼-×-12-inch florist wreath or similar rounded wreath

One 2-×-2¼-×-10½-inch florist wreath or similar rounded wreath

One 1-×-2½-×-12-inch beveled wreath or similar rounded wreath

One 6-inch-diameter ball

Five 2-inch-diameter balls

Four 1-inch-diameter balls

Two 1-×-24-inch dowels

One 3-×-6-inch cone

Modeling paste

White acrylic paint

Terra cotta acrylic paint or color of your choice

One set of battery-operated clock works

One set of clock numerals

One set of clock hands (maximum 3½-inch length)

Assorted silk flowers and foliage

Assorted dried materials

Burnt umber antiquing

Palette knife or spatula

Finished Size: Approximately 8 inches deep by 16 inches wide by 46 inches long.

Technique: The Styrofoam is foam covered with gesso and modeling paste to give the texture of

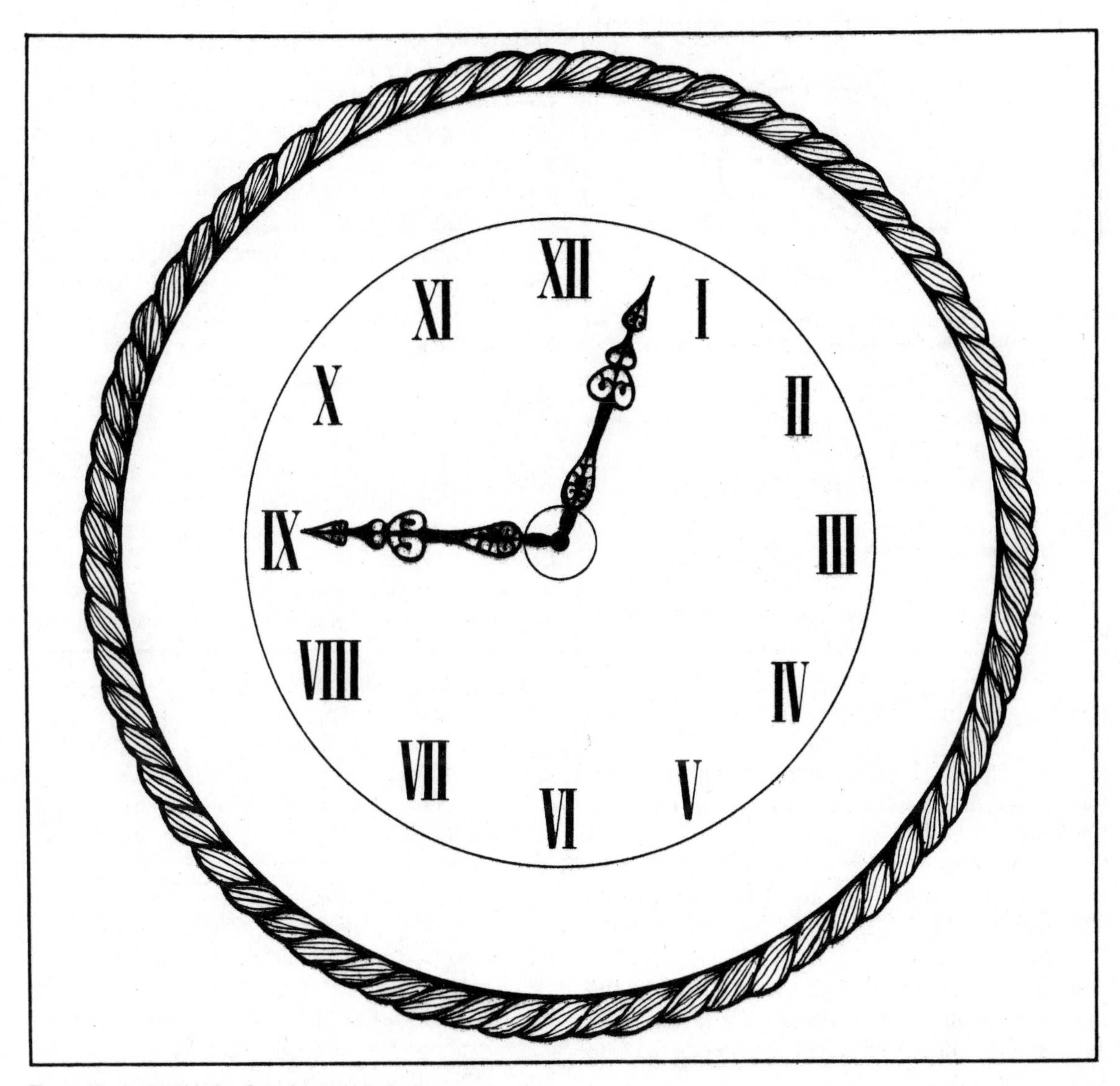

Fig. 6-27. A sketch of a Styrofoam wall clock.

stucco, serving as a foundation for a wall clock.

Preparation: Cut the following pieces of Styrofoam:

□ Cut one 1-×-12-×-36-inch sheet lengthwise to form four 1-×-3-×-36-inch strips.

□ Cut two 8-inch discs in half to form two half circles.

□ Cut the 6-inch ball in half. Then cut each half in half again.

□ Cut the 2-×-2½-×-10½-inch wreath in half to form two half circles.

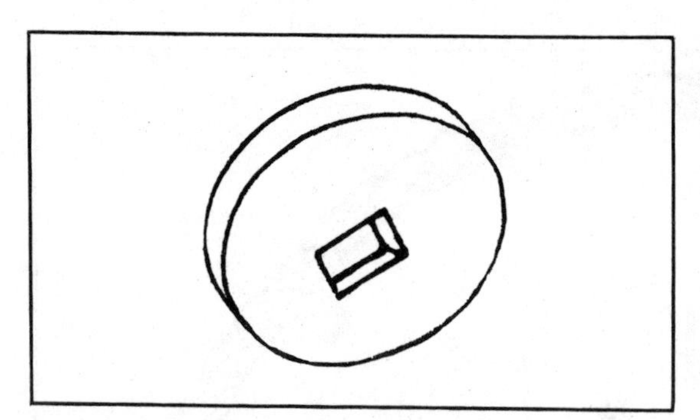

Fig. 6-28. Styrofoam is cut out to accept a battery movement to a depth of 1/2 of an inch.

☐ Cut the 2-×-2¼-×-12-inch wreath in half to form two half circles.

☐ Cut the cone in half, working from top to bottom.

Directions: Refer to Fig. 6-26 for placement of each shape of foam. Place each piece as shown in the diagram on the largest sheet in the following order:

☐ Glue one 1-×-3-×-36-inch strip on top on another strip the same size. Do this with the remaining two strips. You should now have two 2-×-3-×-36-inch strips. Glue one of these on either side of your large sheet as shown on the diagram, keeping back edges even.

☐ Glue each of the two half discs to the top and the bottom of the large sheet as referenced on the diagram. Glue one to extend from the top edge as shown.

☐ Glue the beveled or rounded ring to the center of the sheet, approximately 2 inches from the top of the sheet, and centered between the side pieces. See diagram.

☐ Center each dowel on the side pieces, approximately 6 inches from the top.

☐ Place the quarter pieces of the ball as shown, one at the top, one at the bottom, and one in the center of the sheet.

☐ Glue one half of the 10½-inch wreath in place on top of the quarter ball at the bottom of the sheet. Glue one half of the 12-inch wreath above this, as shown in the diagram. All cut edges should meet the back of the sheet. It might be necessary to pin these pieces in place until dry, using U-shaped pieces of wire.

☐ Glue the cone into place below the center quarter ball. Refer to the diagram.

☐ Glue the 2-inch balls into place; one goes at the bottom of the cone and one at each end of the side panels. Glue the 1-inch balls at the top and bottom of each pole, as in the diagram. You have now finished assembling the foundation, and you should put this aside to dry thoroughly. Pieces should be pinned for reinforcement during the drying process, again using U-shaped pieces of wire.

☐ After the foundation is completely dry, paint the entire surface with a generous layer of gesso. Allow this to dry completely and then cover the entire surface with a thick layer of modeling paste; use a palette knife or spatula. You can texture the surface to look like stucco by making patterns and swirls in the wet paste. Allow it to dry with high and low surfaces.

☐ When the modeling paste is dry, paint the surface with white paint. Let dry. Antique the surface with a coat of terra cotta acrylic paint, thinned with water. Before this is dry, lightly wipe down the surface with a soft rag dipped in paint.

☐ After this coat has dried, add the extra antique look by applying burnt umber antiquing to the crevices. Remove excess with a soft rag.

☐ Allow surface to dry thoroughly. The next step will be to install the clock movement, following instructions on product label for installation. Locate the center point in the back, approximately 9″ down from the top of the sheet and 9″ from the edges. With a knife, create a hole large enough for the stem of the clock movement. With stem in the hole, trace around the movement and set this aside. Working carefully with the knife, hollow out a recess of approximately ½ of an inch into the foam surface. Once the hollow is complete, place the

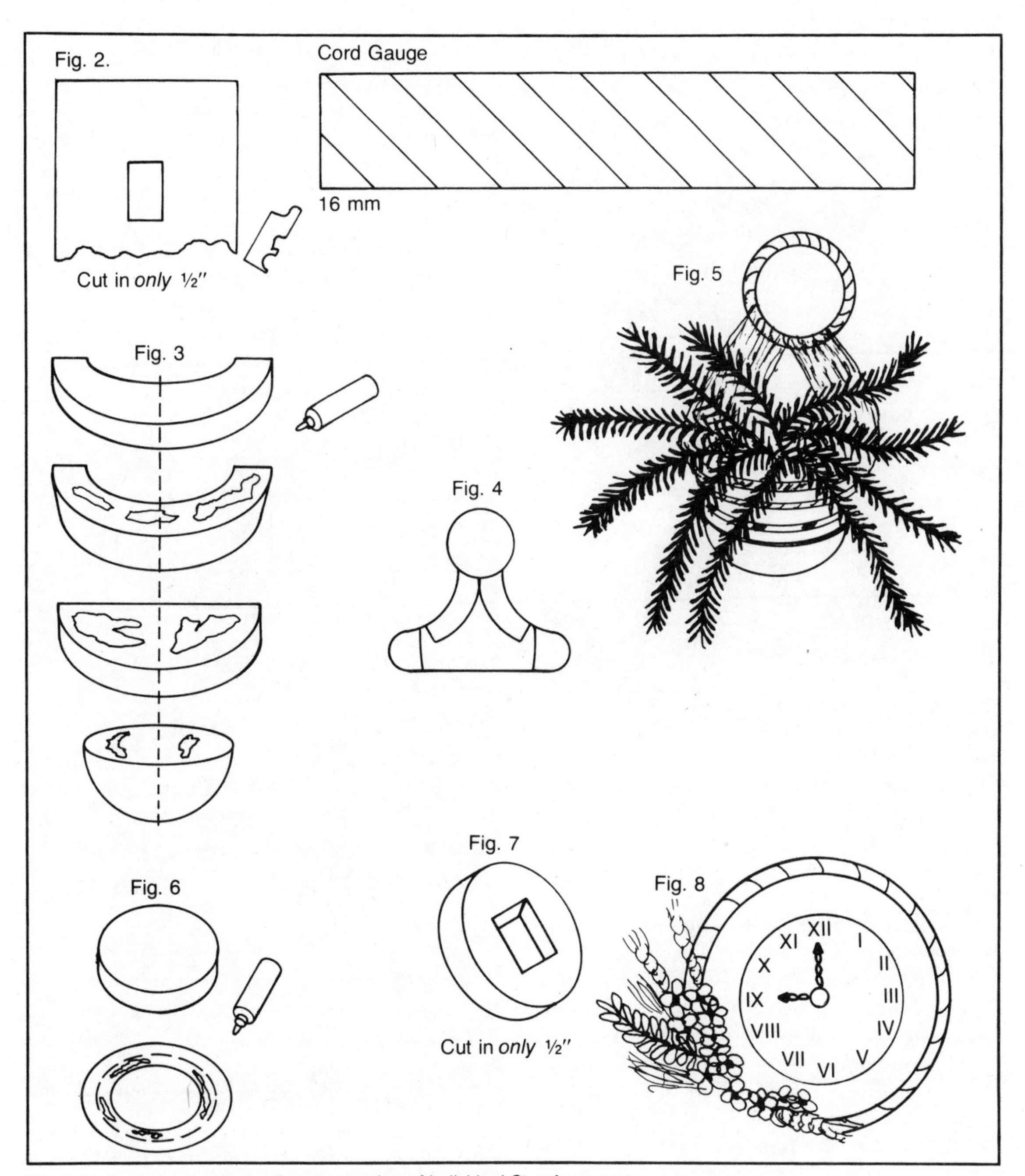

Fig. 6-29. A rough sketch showing sequencing of individual Styrofoam parts.

Fig. 6-30. Styrofoam ball and base designed to accommodate movement with an extra long shaft for the hands.

clockworks inside, with the stem coming out the front.

☐ Following instructions on the label, install the clock hands. This will hold the clockworks securely in the niche. Position the clock numerals.

☐ Set a scrap piece of Styrofoam brand plastic foam in the bowl. Fill the lower bowl with silks and dried flowers and foliage by inserting stems into the Styrofoam. Using foliage, follow this same technique with the upper bowl. Refer to Fig. 6-25 for ideas on placement. Your clock will hang by the hanger provided on the clock movement.

Wall Clock

Materials Needed:

One 1-×-2½-×-12-inch beveled wreath of Styrofoam brand plastic foam or similar rounded wreath

One 8-inch disc of Styrofoam brand plastic foam

One 2-inch ball of Styrofoam brand plastic form

One 12-inch metal ring, gold

One 7-inch metal ring, gold

One ¾-inch metal ring, gold

Battery-operated clock movement

Clock Numerals and Hands

One yard heavy-braid or 16mm macrame cord

Fig. 6-31. Basketball on Wood. Employing half a 6-inch Styrofoam ball. Upholsterer's brass nails were used for hour indications and to mount numerals.

Fig. 6-32. Lantern case housing a beautiful brass ac movement.

Assorted silk flowers, foliage, and dried materials

Antique white or cream acrylic paint

Finished Size: Approximately 12-inch diameter.

Technique: Styrofoam brand plastic foam painted with acrylic paint, serving as base for a clock movement.

Directions:

☐ Glue the 8-inch disc to the back of the 12-inch wreath, centering it in place. Insert wire through the disc and into the wreath to hold secure while drying.

☐ Locate the center point in the back of the disc and form a small hole with a serrated knife. The hole should be large enough to insert clock movement stem. With the stem in the hole, trace around the outside of the movement. Set the movement aside. Working very carefully with the knife, hollow out a recess of approximately ½-inch into the surface of Styrofoam, in which you will place the clock movement. It will be necessary for the stem to protrude from the opening in order to attach the clock hands. Once the hollow is made, glue the clock movement into it and allow it to dry.

☐ Cover the entire surface with a generous coat of gesso and allow to dry thoroughly.

☐ Paint two coats of acrylic over surface, allowing to dry between coats. Let this second coat dry completely.

☐ Glue the 12-inch metal ring so that it rests evenly along the outside edge of the wreath. The 7-inch metal ring should fit close to the inner edge of the wreath, and requires pinning in place in order to dry properly.

☐ Secure nut to stem of clock movement. Glue the ¾-inch metal ring around stem and nut.

☐ After the rings have dried, attach clock numerals and hands.

☐ Apply heavy braid or macrame cord around the outer edge of the project, gluing and pinning as you go. Be sure that the ends come together in the area planned for the floral arrangement.

☐ Carve a small area out of the 2-inch ball to enable you to place it snugly over the wreath. Glue and pin in place at approximately 7 o'clock or the lower lefthand corner.

☐ Arrange florals as desired, inserting stems into the 2-inch ball. Hang the clock from the movement hanger or make a hanger with wire.

Styrofoam Ball and Base

Movements with a shaft for the hands 2⅛ inches long are most unusual. A batch of these turned up

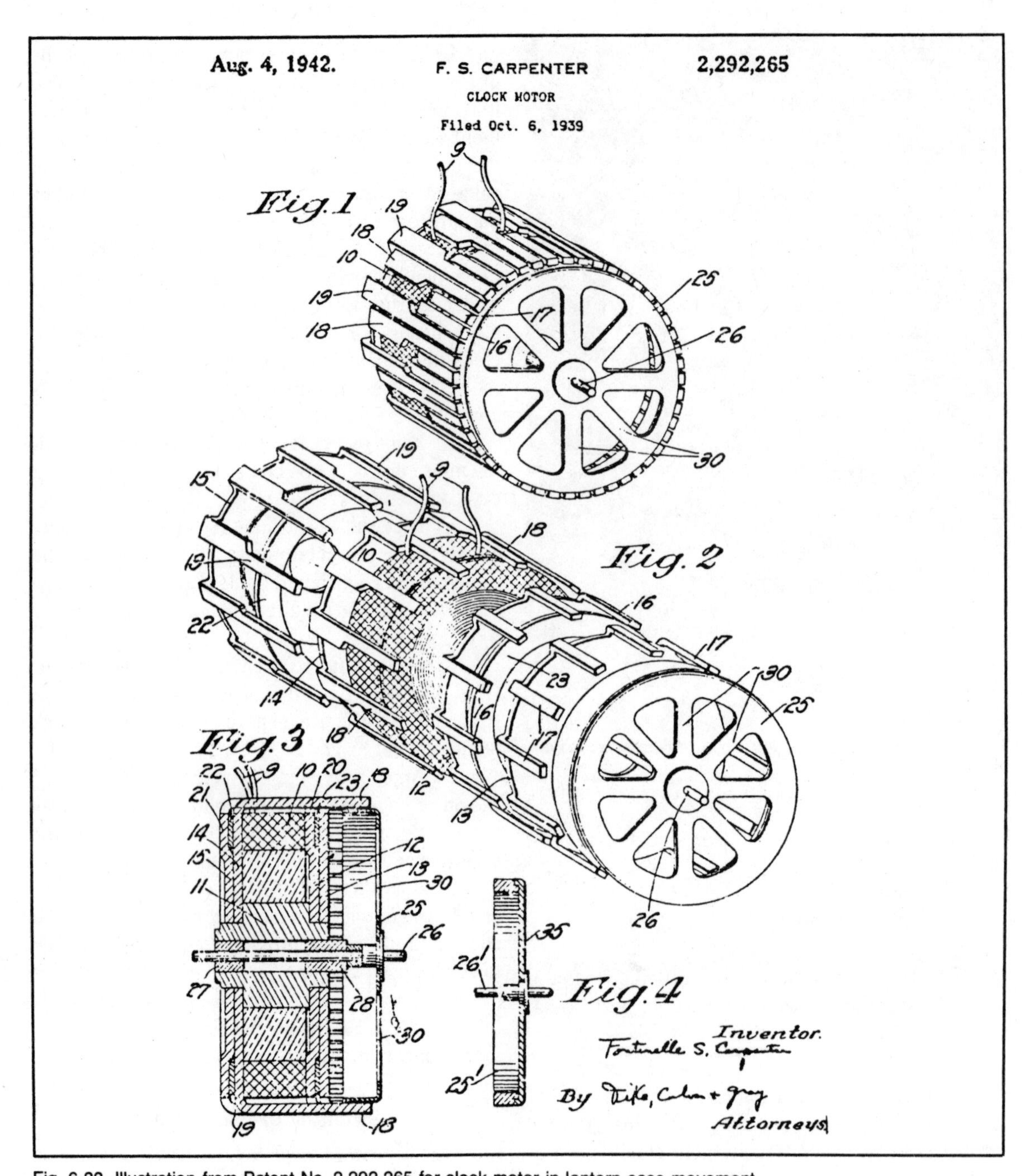

Fig. 6-33. Illustration from Patent No. 2,292,265 for clock motor in lantern case movement.

quite by accident and just how to use them posed an immediate problem. The center threaded-mounting collar extended just ⅜ of an inch from the front of the case while 1⅝ inches of brass arbor protruded from this. These were units quite similar if not identical to, those used in Sunbeam clocks. It was obvious that the movement would have to be positioned some distance from the face of the clock—but how? Here Styrofoam came to the fore. The movement could be buried in a large enough ball. The suitable diameter turned out to be 6 inches and it would have to be done with two balls!

Both balls were cut exactly in half. A circular saw was used for this with the ball being rotated manually as the blade passed through it. A quarter-inch hole was drilled through the center of one half. It was turned over, the shaft placed in the hole, and the contour of the plastic case was traced on the freshly cut portion. One inch of Styrofoam was cut away within the confines of the traced line (the case being used periodically to check the fit). This accomplished, the shaft hole was widened to ⅜ of an inch for a distance of no more than ½ inch in front of the case. This eventually contributed to a tight fit but allowed the arbors freedom to move.

The three hemispheres (one was set aside) were coated liberally all over with Elmer's glue. This step was to prevent the spray paint that was to be applied next from attacking the foam. The pieces were set aside for an hour or so to allow the glue to

Fig. 6-34. Brass and glass case. A small German alarm clock movement with seconds hand and a dial added.

Fig. 6-35. The rear view of a movement in the brass and glass case.

dry. The three pieces were sprayed beige. The one with the cutout and its mate were set aside temporarily and the third was sprayed black, brown, and beige—one after the other. The trick here was to cut a 5-inch hole in a piece of cardboard and place it over the ball half so that about an inch of the piece below the hole would not be sprayed. When this dried, another piece of cardboard with a smaller hole went over the half so that a band of black about an inch wide below it was protected. The brown was applied, allowed to dry, and the procedure was repeated for the final coat of beige. This piece served for the base.

Now the movement was pressed into the hole cut for it. A groove about a quarter of an inch deep was cut from the bottom of the case right out to the end of the foam half. A matching groove was cut in the other half. The leads from the movement were laid in the groove and the two halves were glued together. Care was taken not to glue the leads during the process.

A piece of threaded lamp pipe about 4 inches long was readied for the next step. This is the pipe that normally carries a lamp socket on one end, is put down the center of a lamp base, and fixed there permanently with a nut on the bottom end. This pipe was driven down through the striped base until it exited the bottom and then backed off about ½ inch. A groove about ⅜ of an inch deep was cut from the center hole in the bottom of the base right out to the edge. The leads from the movement were fed down through the pipe and the ball forced down upon it until it touched the base. Elmer's glue was applied between the ball and the base and the two were held together until setting occurred. The leads exiting from the pipe at the bottom of the assembly were crimp-connected to a line cord that was held in the groove with a cardboard disc. A thin piece of felt was added to the cardboard.

As luck would have it, a flat brass ring slightly less than 6 inches in diameter turned up. This was forced open and placed around the ball to conceal the joint of the two halves. Long, brass upholstery nails were used to hold it on. Fancy upholstery nails served to indicate the 3, 6, 9, and 12. Brass hands were fitted and bent slightly to match the contour of the ball. See the following section for the half ball that was set aside.

Basketball on Wood

What to do with the Styrofoam half ball left over from the preceding project turned out to be a relatively minor problem. It was coated with Elmer's glue, allowed to dry, and then sprayed brown. While it dried, a 6⅛-inch circle was drawn on a piece of cardboard. It was divided into 12 segments using a protractor and lines were drawn from the center out through these points. The half ball was set on the circle, and ⅛-inch tape (the kind used in striping automobiles) was laid over the ball using the visible lines on the cardboard as a guide. Once again the half ball was sprayed (this time with Krylon Sunset Orange, #2401).

When it was thoroughly dry, another of the movements was fitted into the back of this half. The procedure for this was the same as above. A wooden disc was picked for mounting and a hole 4 inches in diameter was cut out of the center to allow for hanging the clock and for setting the hands. Bright brass upholstery nails were driven into the wood as hour markers and the plastic numerals were attached with smaller nails of the same type to provide a slightly antiquated look overall.

The wood was cut out at the top to mount the ring, but the clock actually hangs from a little fixture built into the original plastic casing. The hands were fitted in the manner just described.

TECHNIQUES FOR OTHER MATERIALS

It is inevitable that there would be some overlapping when discussing materials from which clock cases are made. In the fit-ups chapter, wood and plastic as well as metal and plastic were combined. Earlier in this chapter, metal and plastic, and wood and Styrofoam are used in combination. That this book deals heavily with plastics should not diminish the importance of the other materials. The intent here is to accent the see-through capabilities of the acrylics in displaying clock movements. Many, many books have dealt with wood, glass, ceramics, and metals in horology. Case work with them will be treated briefly here, and these materials will be discussed more fully in the repair and maintenance chapters.

METALS IN CASE WORK

My experience has been that metal is one of the most difficult materials with which to work unless metalwork is one's particular craft or hobby. It requires more tools than those generally found around the house; some of the tools are required in addition to any used in woodworking. The use of metal for cases was prevalent in the 1800s and in the early 1900s, but one cannot hope to imitate or duplicate the ornate, heavy, bronze-figured mantel

Fig. 6-36. Brass and glass movement (on left) and unassembled alarm clock similar to the one from which it was taken.

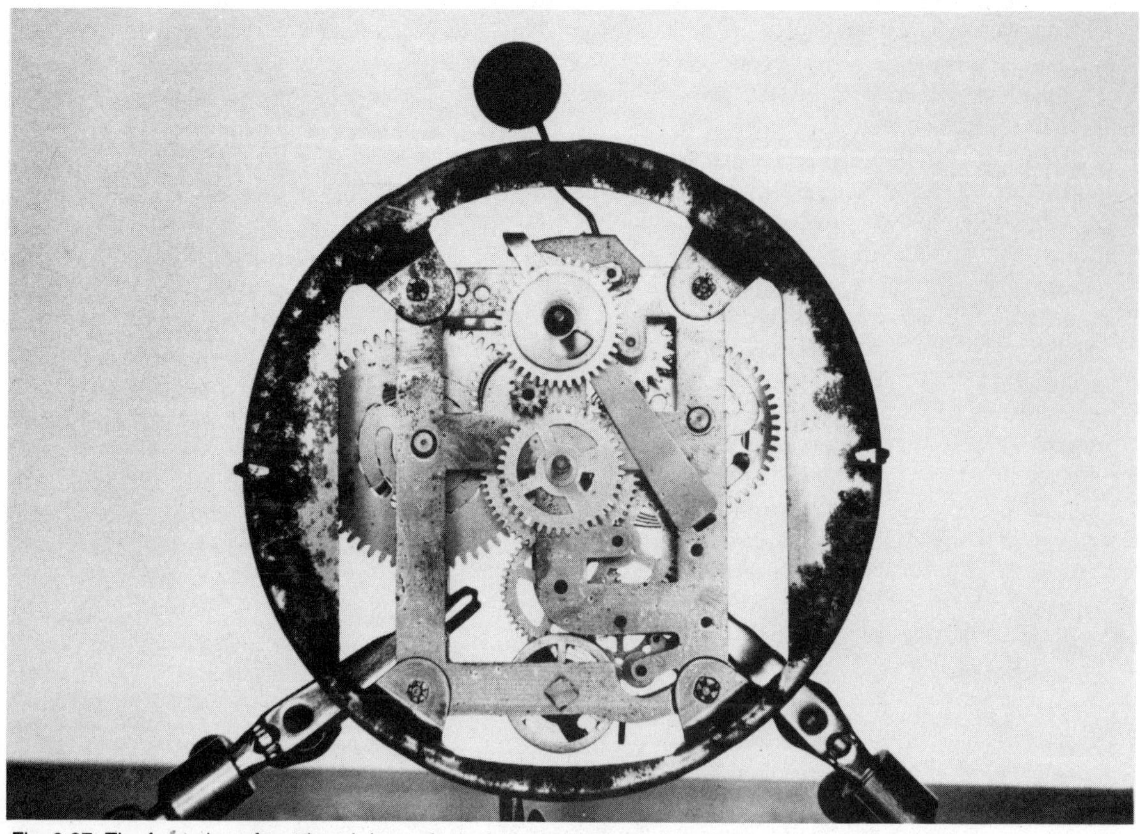

Fig. 6-37. The front view of a stripped down alarm clock showing poor condition.

pieces of that period (and few people would want to). Marble, too, does not lend itself as a medium with which to start from scratch.

For me, one way around this dilemma has been to embrace the metal bases, bezels, pillars, and less elaborate cases from that era when they are in good condition and use them to house or embellish my work. Brass, for example, has escalated in value over the past few years, perhaps because of increasing scarcity. A more pertinent reason, as it applies to horology and clockmaking, is the lasting beauty and the warmth it seems to exude whether it is formed into a doorknob or an anniversary clock. This has been part and parcel of my fascination with older brass movements.

In lieu of being worked up as a case for a piece, metal would tend to become ornamentation or embellishment. Even the devotees of wrought iron display the best of their efforts in the "surround" to a case and in the stand upon which it can be set. Many are the metal cases we've lovingly and painstakingly restored, but few and far between are the ones we've built from scratch. The search goes on always for metal containers that will enhance fabrications, and the one that follows is just such an example.

LANTERN CASE

When I came upon it, the lantern case shown in Fig. 6-32 showed evidence of having served to light the way to someone's front door over a period of many years. It had been painted and repainted countless times. This tended to hide the craftsmanship that had gone into making it. It's sole function had been that of a porch or patio light and, as such, it could have been renovated and pressed into service for another long period of years. This might well have been its function from the moment of acquisition except that just about the same time I discovered a movement that seemed destined to be combined with it. Getting the movement into the housing was what is termed a sleeve job in automotive repair jargon. It would fit only if one had long, thin fingers and a copious amount of patience.

First of all, the movement was one of those rare specimens with the brass plates and wheels that just begged to be preserved for posterity! It was manufactured by the O. B. McClintock Co., Minneapolis, Minnesota. A letter to this company was returned marked "Insufficient Address." The only other clue we had was a patent number that appeared on the back of the movement. The patent office in Washington, D.C. furnished a copy of the

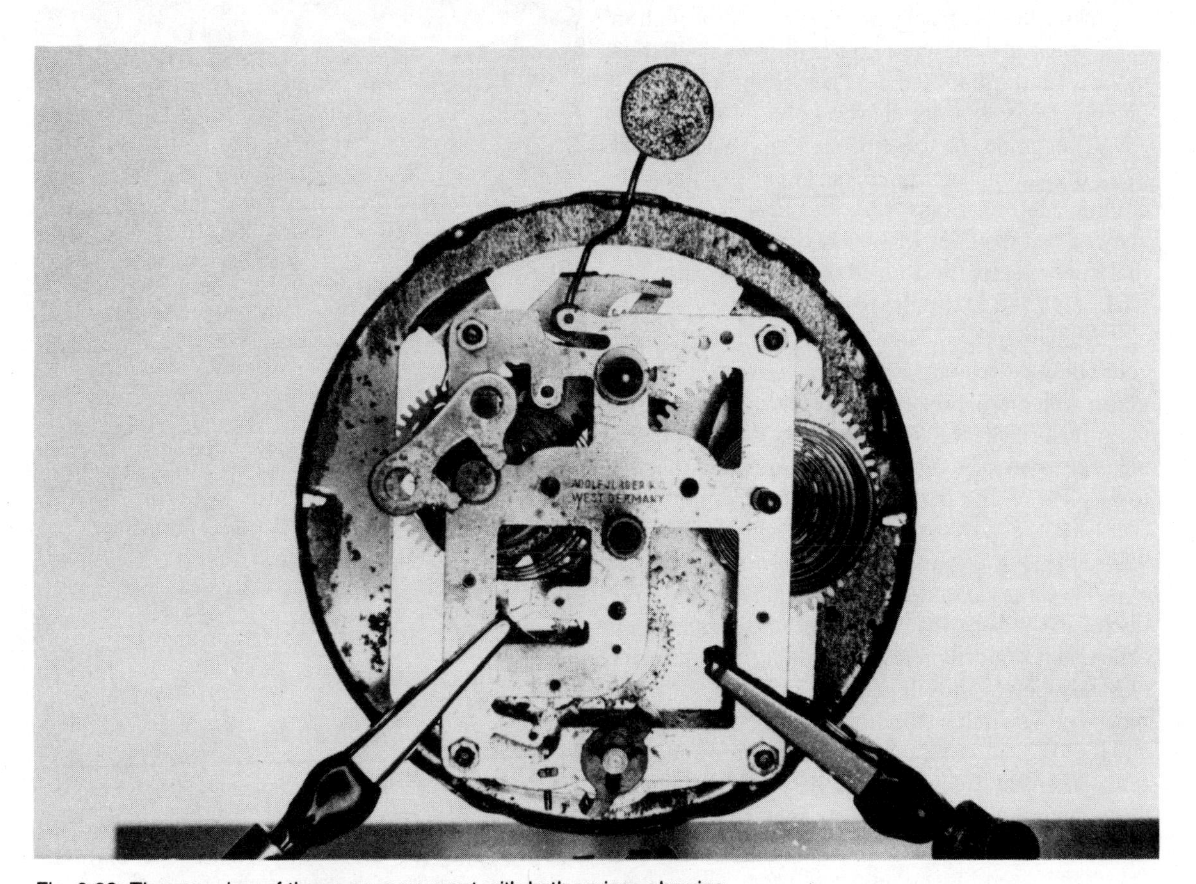

Fig. 6-38. The rear view of the same movement with both springs showing.

patent upon request, and it turned out to be for the little motor that powered the movement. See Fig. 6-33. Invented by one Fontinelle Scott Carpenter, the motor was quite obviously a break-through in the design of small self-starting, single phase, synchronous motors. Many similar motors now appear in clocks by well-known manufacturers, including Seth Thomas, Sessions and Westclox. Carpenter, incidentally, was listed as "Assignor to Waltham Watch Company, Waltham, Mass."

Visible in the photograph is the unusual feature of two fiber wheels, the smaller of the two, meshing with the motor pinion. The movement had been fitted with an alarm mechanism for which power was taken directly from within the motor through an extra lead and a contact set up. I did away with this device, cut the lead off fairly close to the motor, and taped it to prevent its shorting out.

The larger of the fiber wheels made about eight revolutions per minute and I put this to use in adding another brass wheel. Initial inspection showed that the fiber wheel had two damaged teeth that very effectively stopped the whole works. The fix I employed is described in Chapter 8.

Mounting the movement in the "Lantern" case could only be accomplished by removing the front glass, which was broken, and replacing it with one of ⅛-inch-thick acrylic. The glass pieces on the sides also had to be removed temporarily so that the front plate on the movement could be bolted onto the lantern's new outside acrylic plate. This is where the long fingers and patience were required.

The original lantern had a beautiful brass reflector in the back that I polished until it shone by chucking it in a drill press and holding various grits of sandpaper against it until it gleamed. A small light with a switch was mounted high in the housing and the original reflector was used just for decoration. When the light unit and the clock movement were mounted properly, the chore of replacing the side glasses was tackled. They had to be inserted from the bottom, raised high enough to allow their lower end to be put into position, and then held in place by rotating spring fasteners that came down over the top edge. This took much consummate skill with a pair of long, milliner's pliers, tweezers, and a screwdriver or two.

The function of telling time here takes second place to the display of the brass and the movement of the train. No numbers, dots or batons were used

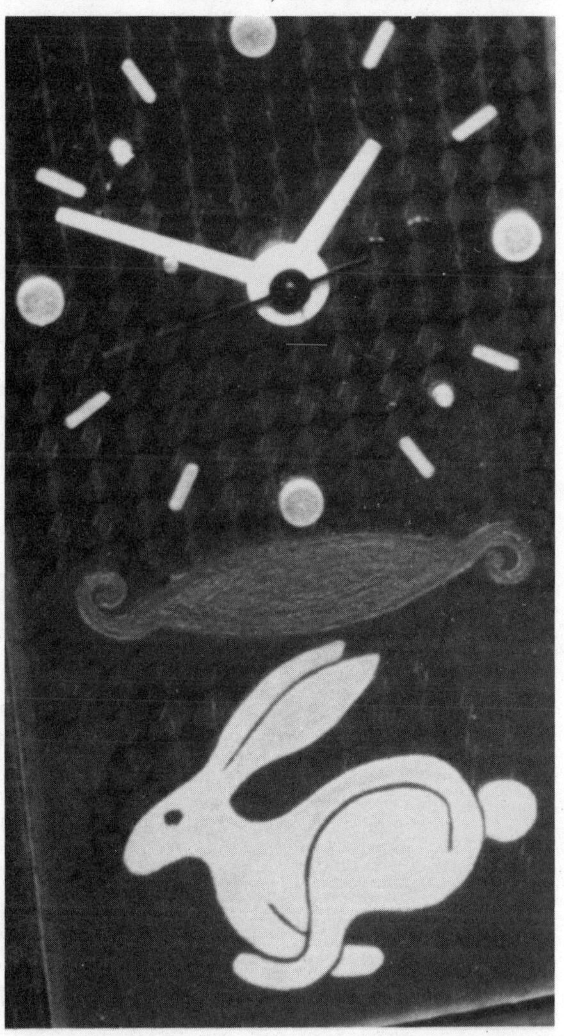

Fig. 6-39. An acrylic/wood combination with an engraved Volkswagen Rabbit on the lower portion of the dial plate.

Fig. 6-40. A desk caddy with a clock and nameplate. The movement and nameplate show in the mirror reflection behind the caddy.

and would be applied only upon the request of a prospective purchaser. The clock, however, is not for sale.

BRASS AND GLASS CASE

A batch of small German alarm clocks with cases in positively abominable condition fell into my hands. They had been made in West Germany, evidentally in large quantities, and bore the legend "Gallatin" on the face. The movements had brass plates and wheels. The clocks were not, by any means, to be considered collectibles but there was something about them that seemed to call out, "Do something with us!"

Further inspection showed that a wheel to which a second hand could be fitted had the end of the arbor "blanked off." It stopped short of penetrating the paper dial just about where the numeral 6 had been printed. The clocks had provision for setting the alarm with a little hand centered in the numeral 12. The markings to set this hand were punched into the metal of the case in the back.

I had a heavy brass dial and glass left from a beautiful Swiss clock lying around and I chose it to frame this unit. A disc of thin acrylic was cut that could be screwed to the back of the bezel, and to this we mounted first a brass ring with adhesive-backed vinyl numerals and then the "Gallatin" movement.

The blanked-off arbor was intriguing enough to warrant some attention, so a little dial clipped from the face of a discarded American alarm was glued to a spacer and then to the frame so that it didn't interfere with any of the wheels in the movement. The end of the arbor that came through it was fitted with a little hand that was also salvaged from an American clock. Sure enough, when the movement was wound, the little hand ticked off the seconds as the movement must have been designed to do originally.

The whole alarm mechanism, including the hammer, the spring and the necessary wheels had been scrapped so that the movement has an uncluttered though not an unbalanced look about it. The original hands with a fluorescent inset in each of them were retained. The movement, as is usually the case with alarm clocks, runs for just about 12 hours. With no alarm capability, its beauty belies the fact that it has no practical use except, perhaps, as a conversation piece on some executive's desk. See Figs. 6-34 through 6-38.

Whether the originals of these clocks sold well or if the purchase price was reasonable, I have no way of knowing. How long they ran before having to be discarded is an unanswerable question also because, for one thing, the pin that normally held the end of the hair spring was placed in a vertical position. This indicates that it might drop out and cause

Fig. 6-41. A propeller hub mounting. The small ac movement fitted into the enlarged hole in the center.

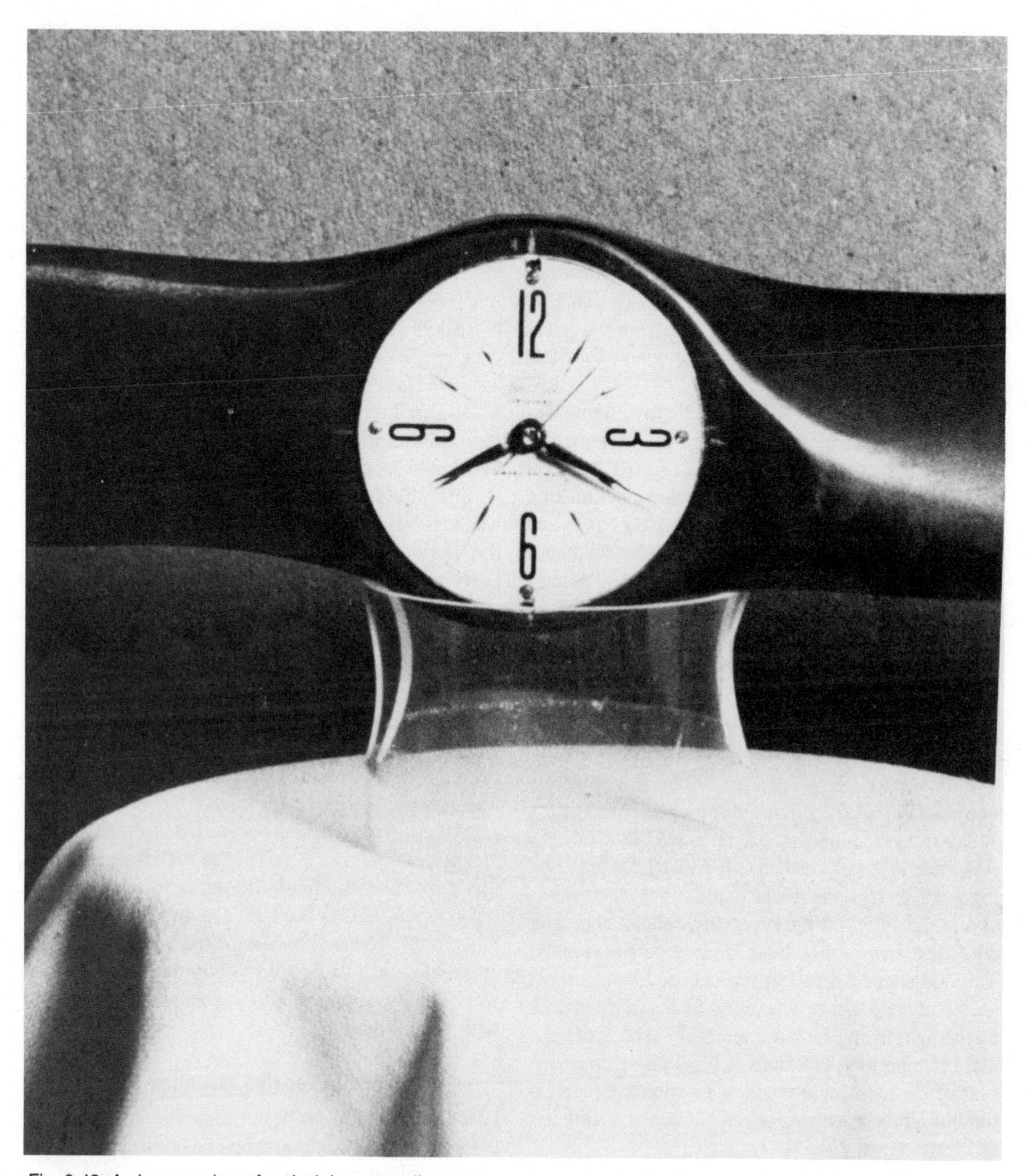

Fig. 6-42. A close-up view of a clock in a propeller.

an immediate stoppage. One specimen had this pin missing and I removed the back. More about this under the heading of Alarm Clocks later on.

WOOD IN CASES

Wood lovers with an interest in horology usually fall into two categories: those who enjoy making the cases for Grandfather, Mantel or Wall clocks, and those who take a slab of natural wood and preserve its shape with polishing and with polyurethane or wax. They might cut a disc or round from the trunk or limb of a tree and make this the nucleus for their timepiece.

Many of the clocks with ac movements acquired as collectibles are enjoyed fully frequently because of the challenge the restoration of the wood provides. The interest in the lathe as a clockmaking and repair tool carries over into making bases and rounded backs for timepieces. On the other hand, plans provided for making wooden wall and mantel clocks do lend themselves to reproduction in acrylic, and it is in this direction I have leaned. One example of when wood seemed absolutely right for one of my projects follows.

Acrylic and Wood In Combination

The timepiece shown in Fig. 6-39 was created in conjunction with an article commissioned by Volkswagen for their magazine SMALL WORLD. A number of the clocks had to feature the VW Rabbit logo, and on this one it was engraved on the bottom of the dial plate. Engraving the rabbit, the hour dots, and batons has been described previously. The odd-shaped scroll in the center hides a short length of acrylic that was glued to the plate to pick up the light from a bulb immediately in back of it for dial illumination. The light is blocked off from the rest of the plate by a piece of prismatic paper extending all the way from top to bottom and from left to right,

The case is actually a box made of beautifully finished wood with the grain showing prominently. The bevel in front was retained and the dial plate was beveled to match it. There is no front cover over the hands, although they are set from the rear of the clock. The little knob on the lower right hand exterior of the case is the end of the toggle switch for turning on the light inside of the case. The Mercedes star might be substituted for the rabbit on the lower portion of the dial. Pieces like this one make highly personalized gifts for people who are known as one marque devotees.

Desk Caddy With Clock and Nameplate

This is another piece produced at the same time as the rabbit clock. The original box from which it is made held surgical tools or something that had to do with the field of medicine. The finish on the wood was magnificent. It was just about 2 inches high on the inside and called for the battery-powered movement to be mounted on its side in order for it to fit.

The back of the box was very carefully removed and a large, mail-slot type piece was cut away with a coping saw. The now famous VW logo was cut out of thin wood (also with a coping saw) and glued to the exterior of this piece with Elmer's Professional Carpenter's Wood Glue. The piece was refitted to the box.

The movement was one with a threaded center mount described in detail in the Fit-Ups section of this book. The mirror shown in Fig. 6-40 shows the top opening in the box and the positioning of the movement. The numbers and dots for the clock dial are gold-colored adhesive-backed vinyl. The hands were cut down in length to fit and carefully sprayed with bronze paint.

Propeller Hub Mounting

Using an airplane propeller as a clock case (Figs. 6-41 and 6-42) can almost be considered as a traditional approach done repeatedly over the last 60 years. It has a special meaning to someone enamored with aviation despite the fact that props might be a little more difficult to obtain than they were some years ago. Mention should be made immediately for the benefit of anyone contemplating this particular project that the wood from which propellers are made is extremely *hard*. Mounting a clock movement and dial in this manner is quite difficult. It is advisable that you spend a bit more time in an effort to match the size of the movement with that of the hub hole of the propeller selected.

The propeller used here had been mated to the shaft of a light, four-cylinder engine. Despite the fact that it had not lifted an airplane off the ground in over 30 years, it was in mint condition, even to the condition of the labels applied by the manufacturer (Banks-Maxwell Propeller Co., Fort Worth, Texas). Removing all of the hardware revealed a hole less than 2½ inches in diameter in the center. The thickness of the propeller at that point (equidistant from the outer tips) was 1⅝ inches. There just was no movement available that would drop into that little area. An Ingraham ac movement turned out to be the smallest available, and even this one required that the hole be opened out to a hair under 3 inches.

An old wood-working technique was employed to enlarge the hole. A circle of 5/16-inch holes was drilled just outside the original circumference of the hole and a coping saw was used to cut from one to another all the way around. This dropped a hollow plug right out of the center that allowed room for the movement with its original dial. An added incentive for the selection of this particular movement was that the dial had hour markers shaped like a propeller blade impressed upon it from the back.

The total weight of the finished product was a definite consideration because clocks in this configuration are usually hung over a doorway. It was a pleasant surprise to ascertain that, with a tip-to-tip length of 42 inches, the finished piece weighed less than 4 pounds! Sawing out the hole was the most difficult part of the project because of the hardness of the wood. Mounting the movement and dial was simplicity itself. Four little holes were drilled in the thin aluminum dial plate close to the numerals and round-head brass screws were used to attach it to the propeller.

GLASS AND CERAMIC TECHNIQUES

Glass is almost never used as an entire case. Side and front panels abound, however, and glass was used ever so frequently during the 1800s into the twentieth century in America and Europe. Many double- and triple-deckers had ornate paintings on the lower glass, some with oval or circular holes incorporated in the center so that the pendulum could be viewed. In the early 1900s, the glass on the doors was used to display advertisements in gilt and black paint.

Jerome Chauncey, somewhere around the year 1817, started a whole new trend with what he called his "bronze-looking glass" clock case. This was a shelf clock and it is said that Joseph Ives actually preceded Chauncey in installing brass and iron movements in looking-glass cases.

It takes quite a bit of practice to cut glass and even more practice to drill holes in it. The following includes all the pertinent data I have acquired over the years relative to those techniques.

Despite a somewhat dormant interest in glass as a clockworking medium, a good specimen of glass that appears about the right size for a clock front is something I never toss away. The painted tablets mentioned above have always held a certain amount of fascination for me when I came across a beautiful thick piece recently, I asked Nancy Miller, who did the art work for this book, if she could paint something on it. We wanted a reverse painting, but this was something she had never done and her attempts were unsuccessful. I suggested that she do something on the front—something with an 1800s look; this she was able to do. See Fig. 6-43.

The design we were working up called for a

Fig. 6-43. A painted glass tablet. A plastic dial with batons is held on the glass with same nut that holds the movement. Painting by Nancy Miller.

hole to be drilled in the center of the unpainted portion of the glass. Although I had drilled a number of holes in plate glass and in bottle glass, the thickness of this particular piece seemed to warrant some practice attempts in similar pieces before this important attempt was made.

I was aware of the technique employing a piece of a triangular file in a dam filled with kerosene above the spot where the hole is to be drilled. Core drills—brass or copper tubes with one end beveled—are used to bite into the glass when the tube is rotated slowly in a drill press. I had successfully used the technique requiring an ordinary masonry drill bit that has a flat piece of tungsten carbide on the end of it.

This time our attempts were a dismal failure. The bit went almost all the way through the sample piece of glass when suddenly there was a sharp CRACK. With the sound, a line radiating from the hole out to the edge of the glass appeared.

Yes, it had cracked and was useless. I did not have one of the spear-pointed carbide drill bits available at the time. They usually have to be ordered by mail so I did the safest thing I could. I took our precious piece of glass with the printing on it to a glazier who agreed to make a hole for us for $8 (and with no guarantee of success). Fortunately, he did it.

His trick, he informed us, lay in the use of special drill bits that, in the larger sizes, cost as much as $250.

"Oh well," we thought, "its back to acrylics for us!"

The masonry drill bit works wonders with ceramics, and that might offer some form of comfort. The lesson learned from the preceding incident is that the thicker the glass is the more difficult to drill it's apt to be. We would have been better off to take the glass circle cutter offered by Brookstone, Peterborough, New Hampshire and make a hole large enough to admit a push-in movement. This also has its drawbacks in the final securing of the movement.

Glass can be painted on, etched and even engraved, but the result is much more certain if it is used as a door or as side panels.

The ceramics is in the form of dishes shaped like the one of plastic mentioned in the Fit-Ups section of this book. There are molded ceramic figures made especially for mounting clock movements, but they also fall in the category of fit-ups. I have never used them.

Chapter 7

Original Movements

DELVING INTO THE RECESS OF CLOCK MOVEments and, more importantly, rearranging the wheels and other vital parts, should always be considered a precise science. Tales abound of people who, with no previous training or experience, have disassembled and reassembled clocks and watches and proudly exhibited them in working condition. After years of living constantly with timepieces, I accept such stories with a skeptical attitude best summed up by the phrase seeing is believing!

Consider the pitfalls inherent in casually prying apart the plates of a spring-wound mantel clock. A main spring wildly ricocheting off the walls of a room would inform one immediately that such energy ought to be released while the spring is still captive between the plates. A little research would clue one as to the correct procedure to be used in releasing it. One would learn quickly about *clicks* and how to allow the spring to unwind slowly while the click was held away from the ratchet wheel. One would also learn how to bind the spring properly so that disassembly could be carried on *without* the spring uncoiling. This information is included in this book and should be searched out and attempted with all due caution.

A spring "letting go" suddenly while still held within the plates is capable of wreaking havoc in what would appear to be a rather confined space. This did happen to me once in spite of all the care I could exercise. In an instant, the movement was ruined; one of the larger wheels in the train lost eight teeth even as other parts were hopelessly bent. Fortunately, the piece was an inexpensive old alarm clock. This did not reduce the shock of observing a precisely made object become instant junk!

That this tragedy did not occur more frequently is due, in part, to the fact that it was alternating-current movements with which we first became involved. I have always had a healthy respect for electricity and some little knowledge of it. I made certain that 110 volts did not do to our person what a heavy main spring might have, had there been one.

Movements by Seth Thomas and Sessions caught my attention initially because the first ones I discovered had brass work that I felt was beauty personified. The first Seth Thomas movement that we modified was such a success that I spent hours trying to find similar ones. I even took a trip to Thomaston, Connecticut only to discover that a flood had destroyed all their records and that the movement was no longer made.

Carrying a Polaroid photograph of the reworked clock, I once stood chatting with an elderly clockmaker of German origin in Red Bank, New Jersey. He examined the photography carefully and then said, "Seth Thomas used to make a movement like that years ago. The only places you might find one are Flea Markets and rummage sales."

To date, I have found four identical movements. One came out of a clock with the legend, WESTCLOX, on the dial!

Our continuing interest in and fascination with these movements is generated by a number of features they possess. The wheels are all brass and large for the size of the movement and, with the exception of those that make up the motion train, all are spoked similar to those in an alarm clock. There is an alarm train fitted that is set from the back (with

Fig. 7-1. A Seth Thomas movement with brass wheel (teeth removed) cemented to fiber wheel (see arrow). Wheels at lower left are added just to cover the unsightly lower portion of the motor.

Fig. 7-2. Another Seth Thomas movement with teeth retained on added brass wheel (see arrow).

no indication of it on the front dial whatsoever). I always discard this portion of the movement.

The clocks with this movement must have been fairly expensive in their day. One I acquired and did not cannabalize was a leather case about 5 inches square with elaborate brass trim one-quarter of an inch wide all around the edge of the front. It has a brass bezel 3½ inches in diameter and fairly heavy brass feet that extend from front to back on both sides. Missing, when we acquired it, were the minute hand and the glass—both of which were replaced. It is in good running condition. See Figs. 7-1 and 7-2.

The motor (self-starting) was made by General Time Instruments Corp. under Patent No. 1935208 - 2015042, which dates it somewhere between 1935 and 1937. A call to General Time elicited the information that the motor was no longer manufactured.

Some movements have the hand-set "stem" protruding from the top while others have it emanating from the back. The latter is preferable when converting them to skeleton clocks.

Figure 7-3 shows three movements: The Seth Thomas described above on the left, a Westclox in the center, and a Hammond on the right. The Hammond is a non self-starting movement and, I feel, it is of increasing value when kept in its original con-

Fig. 7-3. Three movements with potential for conversion to skeleton clocks. Left: Seth Thomas, Center: Westclox, right: Hammond, a bit more difficult to convert.

dition. The Westclox motion train is on top of the plate and the sweep second hand is (on this movement) positioned between the dial and the hour hand. This configuration reveals a wheel making one revolution per minute when the dial is removed. This is visibly more exciting than is the case with a Telechron movement of comparable age, for example. General Electric has movements with the hands arranged this way, but it is accomplished with a series of small wheels added to the movement in a manner that diminishes the visual appeal I seek.

Much of the following information, while compiled under the Seth Thomas heading, applies to ac-movement conversion in general. Different plates, front and back, require different approaches to duplication in acrylic, but their removal will be quite the same. The first clocks I made in this manner had the front plate cut away partially. This was done in order to retain a portion that had a pinion or wheel "pinned" to it. Attaching a pinion to an acrylic plate is difficult in that the acrylic is generally thicker than the plate it replaces, and this can foul up the proper meshing of the teeth which the pinion engages. The ways around this little problem are explained within these pages.

ORIGINAL AC MOVEMENTS

The plates in these movements are normally separated by pillars—short lengths of brass, steel, aluminum or plastic rod. The back plate is removed easily because the rod is threaded on its back end, put through the plate, and secured with a nut. It is sometimes fitted with an internal thread on this end, and in that case a flat-headed screw is put through the plate and into this threaded area. Up front, a thinned length of the pillar is inserted through a small hole in the plate and the exposed end is hit with a punch in such a manner that some of the metal around the circumference of this is pressed flat against the plate. The pillar thus is mounted firmly and permanently. The number of pillars varies from movement to movement. The Seth Thomas movement I've discussed has four; Telechrons most frequently have three, with two extra to hold the unit in the plate. Spartus movements of recent manufacture have three molded onto the front plate, which is plastic. The more movements one examines the more variations in mounting one will find.

Disassembly is a matter of removing the nuts, lifting off the back plate, and carefully removing all of the wheels and other components. Remember, there is no main spring to lash out at you. If the motor is mounted behind the back plate, as in the case of GE Telechrons, it is usually with two additional pillars with recessed threads. The Seth Thomas motor shown in Figs. 7-3 and 7-4 has its motor mounted to the cutaway back plate with two screws, one of which goes through the plate and into one of the pillars. One additional screw goes directly to a pillar fixed to the front plate.

Fig. 7-4. Seth Thomas ac movement mounted in an acrylic case with a brass chapter ring. Owner: Marc Behr.

After removing the back plate, the motor and all the other components the front plate is cleaned with a little alcohol and then little circles are drawn on it with a Chinamarker around the slightly projecting ends of the pillars and around all the holes for the pivots and for the center shaft. Now the ends of the pillars are ground off flush with the surface of the plate. A sharp twist or wiggle with a broad pliers should pop the pillar out without damaging the plate. The marked holes will be duplicated in the new plate.

The motor pivot on the Seth Thomas movement is sheathed in brass and this little shaft will extend up through the new plate. The plate is made of ⅛-inch clear acrylic, 5 inches square. This will facilitate mounting in just about any size wall or mantel clock. Because all of the wheels are visible, they should be subjected to a 15-minute bath in a clock-cleaning solution, dried off, and then polished. Set them aside temporarily and use the original metal plate as a template to make the pivot and other necessary holes. This metal plate should be placed on the back of the new plate and fixed there with double-sided tape. Select clock and watchmaker's drill bits that are each a little smaller than the hole to be drilled for the pivot holes. Each one is carefully broached out (enlarged with a small file) to the proper size.

When all the alarm movement is removed there are seven wheels, including the fiber wheel remaining to be housed under the new plate. They are all on the right side of the movement that exposes the motor on the left. Nothing can be done with ease beneath the top plate to hide the unsightliness of the motor, but wheels can be added above it (deriving their power from the motion train). Positioning the pivots for these additions must be done very carefully. The trick here is to add a pinion with the same diameter and number of teeth as the one that turns the minute hand wheel directly across from it.

A wheel of the proper size is fixed above the pinion. This conceals that the minute hand wheel turns the whole assembly. The position for the new pivot hole is determined by measuring the distance from the center of the minute hand arbor to the center of the pinion and arbor that turns it with a divider. The new wheel will cover just the lower half of the motor so another wheel is added to balance out the assembly visually. While the distance from arbor to arbor must be the same, the final wheel position might vary slightly on each movement that is modified in this manner. The added wheels do not contribute to the accuracy or the performance. They are solely to hide the motor and to add visual interest and excitement.

The fiber wheel (power takeoff from the motor pinion) spins merrily—fast enough to blur the teeth—but to dramatize this movement I resorted to several ploys. I epoxy a brass wheel from an alarm clock with a slightly smaller diameter right on top of it. These should be spoked, but not solid (although the teeth may be ground off by chucking them in a rotary tool and holding the spinning disc against a file). If brass wheels the right size are not available, four evenly spaced holes can be drilled into the fiber wheel or a series of brass dots can be sprayed on it. Two cardinal rules to remember are:

☐ The pivot holes in the top plate *must* match those in the bottom plate or a wheel will wobble, if not jam up the movement.

☐ Pinions drive wheels here. Reverse this procedure and the teeth of the added wheel will skid off the leaves of the pinion—something that is visible immediately.

If the movement is mounted in its case so that it can be viewed only from immediately in front, the back plate does not always have to be replaced. Build depth into an acrylic case, however, and additional appeal may be achieved by "hanging" the components between two clear plates and adding a mirror or mirrored acrylic so that two views of the movement are obtained (Fig. 7-5).

Emphasis has been placed on this Seth Thomas (or Westclox) movement because of the unusual number of its wheels and their appearance and arrangement. Other movements lend themselves to this skeleton treatment, among them the Lux clock Mfg. Co. movement with 2350 Series Motor (5170 L69), described in the Original Cases section. Good for this type of treatment, too, are earlier models of Sessions movements and, of course, some specimens typified by the movement in the Lantern Clock. Refer also to the clear cased Wag-on-the-Wall, Swivel Dial on Stand, and Cube on Cube case.

The final mounting of the movement is much easier if brass or aluminum pillars of the proper length are substituted for the original steel ones. In some cases, 4-40 machine screws with brass collars between the plates work quite as well as the original equipment pillars and add to the general appearance of the movement. Brass or aluminum pillars can be drilled all the way through and threaded to accept fancy screws on both ends in order to hold both front and back plates.

The Seth Thomas movement, shown on the left in Fig. 7-3 (with the line cord attached to it), has the hand-set stem positioned to exit the top of the case. That limits the positioning of the movement as well as restricting the overall size of the case. This placement of the stem also requires a series of extra wheels to turn the hands. All of them, with the exception of the one meshing with a wheel on the center wheel arbor (the one to which the minute hand is fitted), can be eliminated in two ways. The hands can be set from the front of the clock or the center arbor can be lengthened to protrude through the back plate. The end can then be threaded and fitted with a knurled cap. This arbor can also be extended forward as is the case with the Westclox movement in the center of Fig. 7-3 (used in the Fiber-Optic Basket Clock). In this last event the hands would be exposed permanently. Uncovered hands, it should be noted, are not unusual on modern clocks.

The movement shown on the right in Fig. 7-3 is a non self-starting unit made by the Hammond Instruments Co., Chicago, Illinois, in the 1930s. While these movements might turn up occasionally, they are not recommended for this purpose. They are difficult to work with because the plates are white metal with the pillars split in half and cast as part of the plates. Moreover, they are considered collectibles. If they are discovered with irreparable cases, the movements should be saved for possible use in the restoration of similarly, valuable units.

A little background material is in order here if for no other reason than to pinpoint the date of manufacture of the movement surrounded by its large acrylic dial with "lighted" spandrels. It almost commemorates the end of an era—that of the appearance of craftsmanship in the midst of mass production. Brass Wheels! Brass Plates! Brass Pillars! All of these in electric movements join similar attributes in mechanical (spring-wound) movements in the realm of horological history.

The movement in question is a Sessions (Fig. 7-5). This company evolved from the famous E. N. Welch Manufacturing Company that made clocks for more than 85 years. Somewhere shortly after 1900 one of those company takeovers, so familiar today, occurred through the manipulation of Welch stock. William E. Sessions, aided and abetted by his nephew, A. L. Sessions, wound up president and treasurer of the E. N. Welch Mfg. Co. They put a lot of money into revitalizing the company, and changed its name to the Sessions Clock Company. Here again is an example of someone who bought into a clock company with the result that the timepieces became legend while bearing his name.

In the mid-1950s, the company name was shortened to the Sessions Company, only to be bought up by the Consolidated Electronics Industries Corporation of New York. All of the Sessions' clockmaking paraphernalia was sold off to a firm in Brooklyn, N.Y. in 1968.

Fig. 7-5. Sessions movement on acrylic plate with engraved spandrels. The fluorescent bulb in the base illuminates the faceplate. The on/off switch is on the left.

It becomes apparent, therefore, that this movement was manufactured some time between 1956 and 1960. There was another Sessions in the clock business during this period, but his company was known as the New England Clock Company and the movements used in the clocks made by this Connecticut company were purchased outside.

The movement seems to come from an even earlier period with its spoked wheels and bright nut fastenings. I decided to keep as much of the original movement as possible. To expose the "going" wheels, the top plate was cut away with a coping saw. The space between the hands and the first wheel in the motion train would not permit the installation of a mounting plate. Therefore, the movement was attached to the dial plate by bolting it to three lugs glued to the plate, the center of which had been cut out. Light from the fluorescent fixture in the base was piped up through the plate and formed a visual halo, illuminating the movement.

Viewed head on, without the light, just the brass and the moving wheels catch the eye. Push the switch on the left of the light housing and the spandrels appear suddenly. The dial plate is 9 inches long by 7 inches high and the whole clock is at its best when set in a dark area or one with a neutral background. It is narrow enough from front

Fig. 7-6. A modern wag-on-the-wall. The spring-wound movement of Korean manufacture runs 30 days.

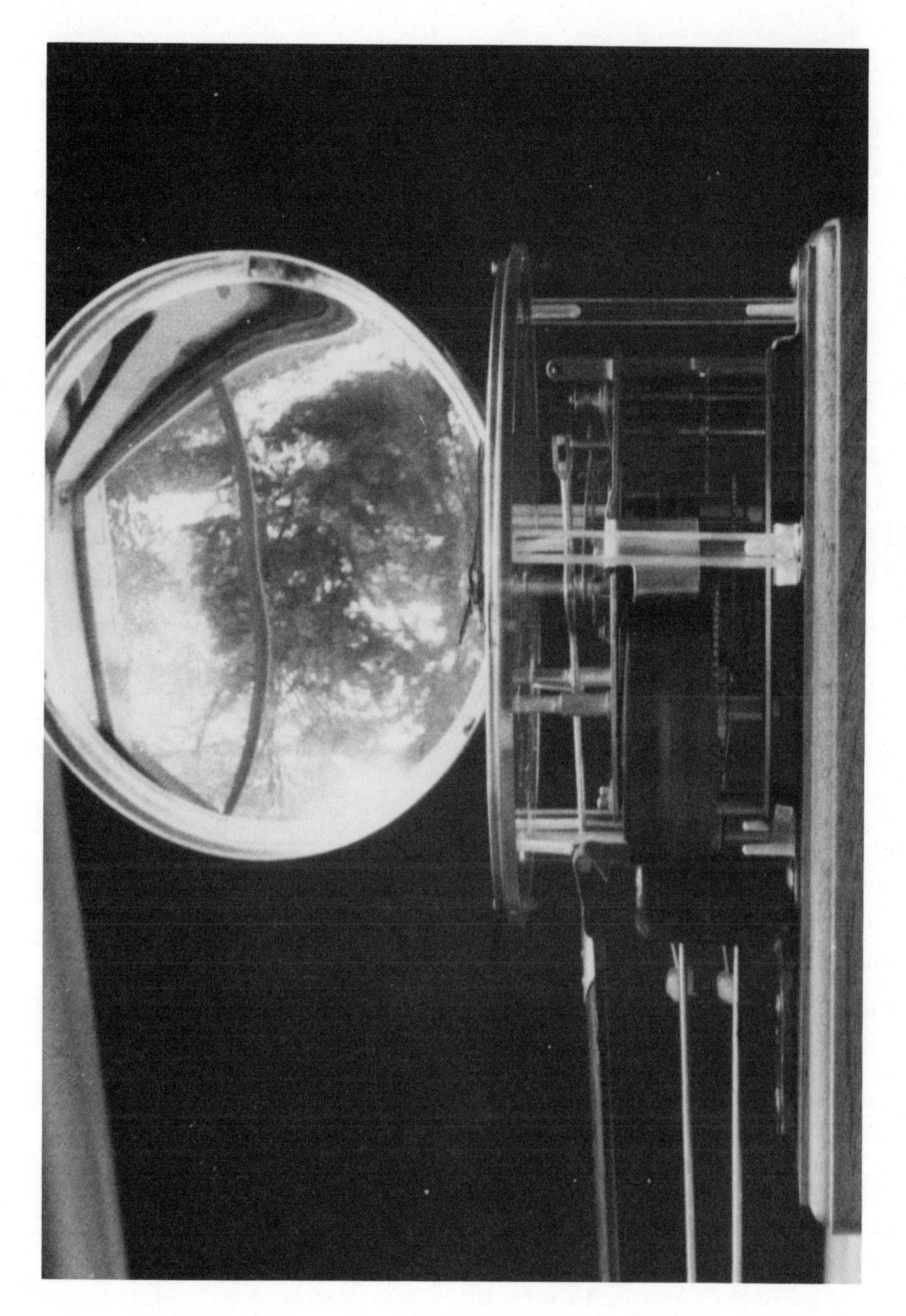

Fig. 7-7. A side view of a wag-on-the-wall showing fabricated acrylic pillars that hold the chapter ring and glass door with a brass bezel to the movement.

to back to be set on a bookshelf and still be easy to read if books with dark bindings are lined up behind it.

KEY WIND MOVEMENTS

Spring-powered movements do not lend themselves to the type of modification explained in the foregoing passages. Many of them are fitted with brass plates and movements (Figs. 7-6 and 7-7). But unless they are 8-day or 30-day movements, they require too much attention in that they must be wound daily. The clock described under the heading Brass and Glass exemplifies what can be done with the 30-hour movements, but only one other specimen I assembled justifies mention here.

MODERN WAG-ON-THE-WALL

The following advertisement appeared in the classified section of a suburban newspaper just outside of New York City: Clock movements, new, thirty-day, hour strike with pendulum and key, $10.00. Call (a local phone number).

It developed that an elderly retired gentlemen who tinkered with clocks in his basement had three movements he wanted to dispose of. After a cursory inspection, I bought one. It had 5-inch-diameter round brass plates, a pair of tuned rods, and a 12-inch pendulum rod. The bob was 3½ inches in diameter. As a shelf or wall clock, it would require a case at least 2 feet high to be enclosed completely.

I decided to make a wag-on-the-wall, as was done as far back as the early 1700s in America. In those days, clockmakers frequently made and marketed just the "works" alone. If the prospective buyer was wealthy enough, he sought out a cabinetmaker and ordered a case. Cases were expensive and, because the clockmaker provided a dial or face, many buyers simply hung the new clock up on the wall au naturel—without any casing whatsoever.

This movement, made in Taiwan, did not have a dial or a backboard, and that presented a minor problem. Another inspection of the movement revealed that the pendulum was hung in front of the rack and snail. The wire crutch had a big bend in it to the left of the hand shaft, permitting it to swing back and forth with complete freedom. It seemed such a shame to cover everything up. Fortune would have it that a 7-inch brass skeleton chapter ring, complete with a round glass door, was available and I put it to use.

An acrylic disc about 7¼ inches in diameter was cut and the chapter ring fitted to it with tiny brass screws. A beautiful mahogany board, originally intended for use as a plaque, was selected for the backboard. It measured 7×9 inches. That was large enough to accept the lugs on the back of the movement despite a ¼-inch ogee molding that had been cut around the leading edge.

The lugs (mounting brackets) had two holes in the portion that fit against the board. One was used to screw the movement to the backboard and the other for ¼-inch square acrylic pillars to mount the disc that held the chapter ring. The pillars were flat on the outer end and had 6-32 machine screws threaded into the other end. Four holes were drilled into the board and threaded. Each of the pillars were screwed, through the lugs, into the board. The disc with chapter ring was positioned on the other upright end of these pillars and cemented permanently in place.

All that remained was to add the hands and mount the chime rod assembly beneath the movement. Two hammers hung down to strike them on the hour. The modern version of a wag-on-the-wall was then hung where it was to be used and the springs for the two trains wound through holes previously drilled in the acrylic disc. A bit of horological trivia that might prove interesting applies here. The holes for the winding arbors on a 30-day movement are usually just below the numerals 4 and 8 on the dial, and just above these numerals on an 8-day movement.

Emphasis has been placed throughout this book on the fact that prized clocks should be cleaned every two or three years. This rule-of-thumb applies to the wag-on-the-wall because the movement is exposed on the sides. Fitting a cylindrical shroud to enclose it is difficult because of the pendulum and the chime rods. The shape of the case has quite a lot to do with the sound of chimes. Placement in the home (forget the kitchen!) and ambient room temperature also have a direct bearing on how often a clock must be cleaned or maintained.

There is a paradox in the dearth of key winds movements discussed here because I am totally enamored of the sound of ticking timepieces. A portion of this can be attributed to an on-going interest in older, battery-powered movements that have a healthy tick that can be magnified by the casing if desired. A discussion of them follows shortly, after this final comment on key winds:

> There was a man, he had a clock
> his name was Mister Mears,
> And every night, he wound that clock
> for five and forty years.
> And when at last that clock turned out
> an eight-day clock to be,
> A madder man than Mister Mears,
> I never hope to see.
>
> Traditional

MODIFIED BATTERY-POWERED MOVEMENTS

So many, many of the modifications made to the alternating-current movement are adaptable to the battery-powered movement! The decided advantage with the latter is that no line cord is needed. Additionally, there is no sudden stoppage in the event of power failure. The biggest disadvantage is that the wheels, except perhaps in the older movements, are much smaller. This would tend to restrict any models made into skeleton displays to desk or table clocks that are viewed from relatively close distances. Dials for movements used normally, it has been shown, can be just about any size.

Initial counteraction of the size limitation would be to use the largest of this type of movement available—big old ones almost always fitted with D-cell type batteries. See Fig. 7-8. Many of this type are still readily available because they were produced in large quantities by a number of manufacturers in the United States and overseas. Some examples, shown in Figs. 7-9 through 7-14 include, Bayard of France; Schatz of Germany; Cosmo Electric of Japan, and Ingraham made in the United States. There are numerous others, of course, but work with these is much closer to the ac- and spring-wound offerings presented in preceding sections of this book.

Working with the more modern, smaller units manufactured today approximates watchmaking. There is a great deal of satisfaction to be derived here as one becomes better acquainted with the use of the jeweler's loupe and the manipulation of tiny tweezers and screwdrivers. Please note that battery-powered movements have direct-current (dc) motors! Of prime interest is that, while most of these movements are of the threaded center shaft fit-up variety, this capability might have to be forfeited on occasion to the skeleton appearance that is among the most significant effects to be achieved.

The differences in the various manufacturers' approach to getting energy from the D-cell battery to the movement is fascinating, yet they are not totally divorced from each other. Each manufacturer chose a particular route prescribed by firmly established electronic principles quite worthy of a brief discussion here.

The little discourse that follows must be construed as but the tip of the electronic timepiece iceberg and is governed totally by the particular specimens I have used successfully. This excludes "no moving parts" quartz movements.

Fig. 7-8. An older, battery-powered movement powered by a D-cell battery made by Ingraham.

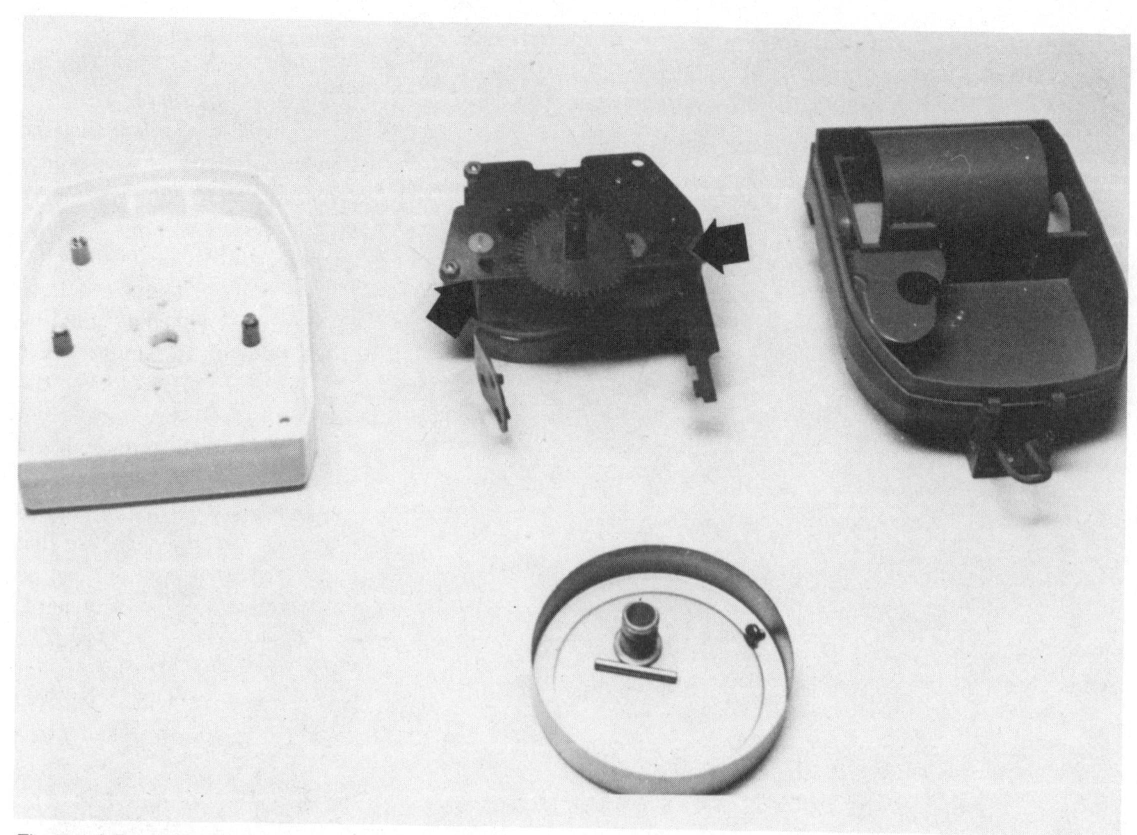

Fig. 7-9. A French Bayard movement with opened casing on either side. Studs are normally mounted in holes indicated by the arrows.

Many of the units with which I worked, and some of which are described in detail in the repair chapter, fall into the category of rewind movements. They are, in essence, mechanical clocks with mainsprings and escapements, including hairspring and balance wheel. The difference is that the mainspring, mounted in a barrel, drives the train and escapement for a period of brief duration (about a minute or so). It carries with it a ratchet wheel and a pawl with a contact fitted to its end. A conventional spring pulls against the pawl and forces the end with the contact toward a matching contact on an armature. When they (the contacts) touch, a circuit from the battery to an armature is closed and the contact on the armature is yanked down, forcing the pawl in the opposite direction. Contact is broken and the spring on the end of the pawl takes over, forcing it forward again while allowing the mainspring to move the train. The train does not stop so quick is the moment of contact when the pawl is shoved back.

In other versions, small motors are activated to perform the rewinding. They use a worm gear that is turned by the little motor for winding, but cannot turn the motor so that all the energy stored in the spring has to be discharged through the train.

Still another variation is the magnetic balance type (used in the Scrap Paper Caddy—Fit-Up section). The balance wheel is generally equipped with one or more little magnets that oscillate just above and below tiny double coils of wire, minus the usual center core. When the magnets swing over the coils, one of them receives a minute induction that it passes on to a transistor.

The second coil returns this current, amplified, as an impulse to the balance. There is a little worm gear on the arbor of the balance that moves a wheel linked to the train. The movement is regulated in quite the same manner as an ordinary alarm clock (by an adjustment of the balance spring). These movements, until the advent of the quartz, were considered quite reliable and accurate. They run easily for a year or longer on one little C-cell battery.

The movements should be disassembled and examined closely. Most of them are still in good supply and most inexpensive when acquired at flea markets, garage, and rummage sales.

The old Ingraham (Fig. 7-8) is very similar to the ac movements elaborated on earlier. The little synchronous motor, made in Japan, powers a stubby shaft with a brass spring on it. The end of the spring is bent out so that as it rotates it hits a pawl fixed to a pinion that turns a fairly large wheel and carries the energy all along the train to the escapement. The pivot holes in the plates are as tiny as those in a watch, the pillars are crimped to the front plate, and the movement in general is just not one of the easiest to modify.

The French Bayard, frequently used in American clocks, is a joy. It is also the most unusual of the lot. Under normal circumstances, if anything goes wrong with a battery-powered movement it is discarded. This one definitely falls into that category because it would be very difficult to reassemble if it were taken apart. A rotating pin from the shaft of the motor activates a pivoted cam, which, with its up and down movement, oscillates a ratchet-toothed wheel on the same arbor as the escape wheel. The hairspring rides beneath the ratchet-toothed wheel. Once the top plate is lifted off, the cam drops down and all the wheels fall over.

All of this is complicated by the fact that the rear pivot for the cam is mounted in a hole in a bridge below the back plate. All kinds of patience, manual dexterity, sleight of hand, and some juggling are required to replace the bridge and all of the wheels. All the holes in the front plate are not there for cosmetic reasons; some of the arbors must be reached through them for proper alignment. As is the case in the reassembly of alarm clocks, the balance wheel should be left for the very last because it can be put in place after everything else is fastened down.

Rather than replace the top plate, I opted to drill it out until it resembled Swiss cheese so that the unusual movement of the pivoted cam can be viewed in the light of how it affects the escapement. A sweep second hand can and should be added. The arbor is there and can be reached by fitting a collar to the center back of the second hand. The big contacts for the battery can be removed and wires

Fig. 7-10. The Bayard movement before any modification.

Fig. 7-11. The French Bayard movement after having been reworked.

can be substituted as shown in Fig. 7-9. Mounting the movement can be accomplished by any of the methods suggested either in the Original Dials or Original Cases sections.

Because of its size, the Bayard movement will make an exciting table or mantel clock modification, but it should be tackled only after you have acquired a bit of experience with some of the other less complicated movements.

Any modification of the top plate will be facilitated if the tops of the two pillars on the right side (they are crimped over the plate) are ground away. This will preclude the necessity of removing them from the bottom plate (an action that drops the bridge beneath it). They can be drilled and threaded

Fig. 7-12. The Bayard movement with the top plate removed. Note that wiring has been substituted for battery contacts.

Fig. 7-13. A Japanese battery movement frequently used in Verichron clocks.

for acceptance of short machine screws when the top plate is to be refitted.

Many American clocks are fitted with foreign movements (more often than not of German or Japanese manufacture). The legend on the dial can be deceiving so it pays to turn a specimen being offered for sale over to examine it in back carefully. Adhesive labels often conceal the fact that the movement is not of American origin.

A very popular name in decorator clocks is VERICHRON. Over the past few years, my examination of more than 50 copies of this brand revealed Japanese movements, some even fitted with true pendulums. Earlier versions of these particular movements, made by Cosmo Electric in Japan, had metal front plates that did not lend themselves to elaborate modification. More recent versions, however, have much of the metal replaced by plastic. This includes pillars, wheels, and bridges. The unit, as it comes from the factory, measures approximately 3×3¼ inches. Much of the original housing can be discarded reducing the overall size to less than 2×3 inches! This includes a plastic cover for the movement. The white plastic and brass in combination is pleasing to the eye and can be enhanced somewhat by judicious removal of portions of the front plate, particularly in the lower right portion, that will reveal the balance wheel and the escapement mechanism with its plastic anchor.

These movements are threaded, center-mount fit-ups. As long as the threaded collar is retained, there should be no problem in affixing them to a new custom-designed dial plate. The back portion of the original housing is snap-fitted into the front portion and then secured with two self-tapping screws. Cutting away about 1/8 of an inch of the clear plastic lugs on either side of the back case will permit its continued use, although there will remain an open area all around the unit unshielded from the atmosphere. A purist might go to the trouble of cobbling up an extension to the clear plastic cover, but it hardly seems worth it.

The overall finished effect will be much more pleasing if the plate is made of 3/8-inch stock. This will reduce the distance the threaded collar and the shafts for the hands extend forward of the face. A thicker plate, remember, lends itself to rear-etching and engraving. Don't even entertain the thought of playing around with adding extra wheels on these units. There just isn't enough uncluttered area for it.

Retention of the battery housing (the lower portion of the original case) is expedient here. The leads from the movement can be lengthened to permit the battery to be housed somewhere out of sight. This ploy also works to maintain the good

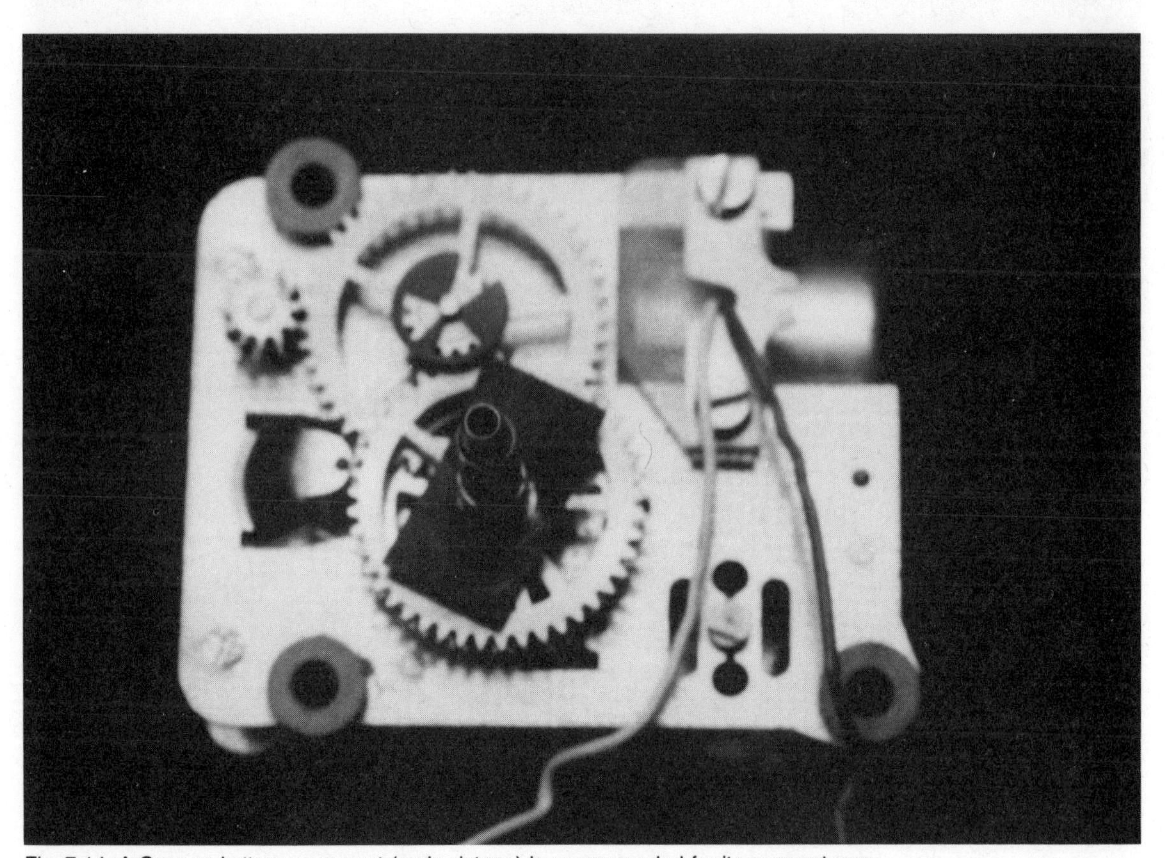

Fig. 7-14. A German battery movement (early vintage) is recommended for its compactness.

contact for the battery terminals offered by the box.

The older German units I have examined and used, on occasion, are frequently of the type employing a little motor with a helical gear on the shaft that turns a wheel to wind the mainspring. They are compact, usually being no wider between the plates than the diameter of the motor housing (about an inch).

When removed from the original housing, the movement measures approximately 2×2¼ inches. Their size makes them ideal for desk caddies or for incorporation into some other display item. The only difficulty is that most of them have metal front plates that require consumate skill with fine tools in small places. I've used them in projects where the appearance of the dial is more important than the movement or what has been done to it and I've used them as reliable, durable replacement units.

The front plate is generally metal with some sort of punched dot detailing across its whole surface. Little or nothing can be done to alter this plate to reveal the working of the escapement except, perhaps, to cut away the lower right-hand portion as described previously. They are attractive when fitted with brass, spoked wheels in the motion train. They are generally threaded, center-mount fit-ups (an original aspect that can be ignored when reworking them). They will mount readily to a new plate, thick or thin, with three or four machine screws or, even more simply, on studs.

The burgeoning production of quartz, LED, and LCD movements will make most of the units described above much harder to acquire in the years to come (which is regretful). It will not be too long before many of them will be considered true collectibles. I've tried to work with the most heavily mass-produced movements that lost popularity because of the new movements cited above or simply because someone tired of the case in which they were housed. I hope that there were enough of them so that others will not be deprived of the joy of collecting for horological or historic reasons.

Chapter 8

Repairs

HUNDREDS OF BOOKS HAVE BEEN WRITTEN about the repair of clocks and watches. Many of them are slanted toward those few in our populace who are dedicated to the preservation and restoration of fine timepieces. Most of them give the lowly, inexpensive alarm clock, the alternating-current electrics, and the battery-powered movements short shrift. No matter how low the purchase price or how huge the quantity mass produced, a great deal of thought and ingenuity went into the original designs, and anyone who wishes to should find at least one tome that will enable him to, as the English say, put one right again.

It is true that much of what appears on shop shelves these days just does not warrant much effort in the event of stoppage what with the prevalence of plastic (small, almost irreparable pivots, etc.), throughout the movements. Yet the challenge of attempting some sort of fix should not be sidestepped.

Horology should not be the exclusive province of wealthy collectors. The sheer excitement to be found in the extremely well-regulated sets of wheels that go to make up a timepiece should be available to all. Admittedly, the subject is so big that one could easily specialize in one segment of it. The basics, once acquired, are such that "touching" an alarm clock is tantamount to opening the door to the solution of all the mysteries in a beautiful French movement. Following immediately are a number of the repairs that can be made without elaborate and expensive clockmakers' tools and that fare well to encourage some readers to venture into areas quite thoroughly covered by the myriad of other books.

Satchel Paige, the famous black baseball player, once said, "Never look back. Something may be gaining on you!" Not true in this case! It is my feeling that all the horological discoveries over the past 400 years are pointed up in the craftsman-

Fig. 8-1. Brass and marble Telechron ac movement after restoration. Owner: Gerri Miller.

ship of the last 80 years or so, and it is this period that is covered herein. The future belongs to the kids in high school and college who are steeping themselves in computers, logarithyms, and advanced electronics. Quite like today's little old watchmaker who is just about to retire, we are not prepared to deal with quartz, tuning forks, watts, and ohms in timepieces.

BRASS AND MARBLE TELECHRON

Mention of Henry Ellis Warren and his patented Telechron is made in the section on Clocks of

Original Design. The marble and brass beauty shown in Figs. 8-1 through 8-9 is a prime example of what made Warren's work so special. When it came into my studio, it obviously had not run for quite some time. The first step in repairing it was to remove the cup in back so that the line cord could be checked thoroughly from the plug right up to where it was soldered onto the coil. The brass cup on the back had to come off. Normally, two screws are removed, the knurled knob on the hand-set stem or arbor unscrewed and the brass cup is removed with the line cord still threaded through a hole in the back of it. The movement glass and bezel is then free to come out of the front of the mounting hole in the marble.

In this case, the knurled knob would not come off! In desperation, I took a bolt cutter to the hand-set arbor and cut through it just beneath the knob. It (the arbor) would have to be replaced, of course, but this was not a serious matter because the Telechron movements are still easy to obtain. The most practical method of repair seemed to be the replacement of the motor. Two screws free it from the assembly so that a substitution can be made. This is generally the case with most if not all of the ac clocks of that period.

Because the clock was to be put in good running order it had to be cleaned. The bezel was removed by lifting the tabs. It and the glass were set aside and the hands were removed. Care must

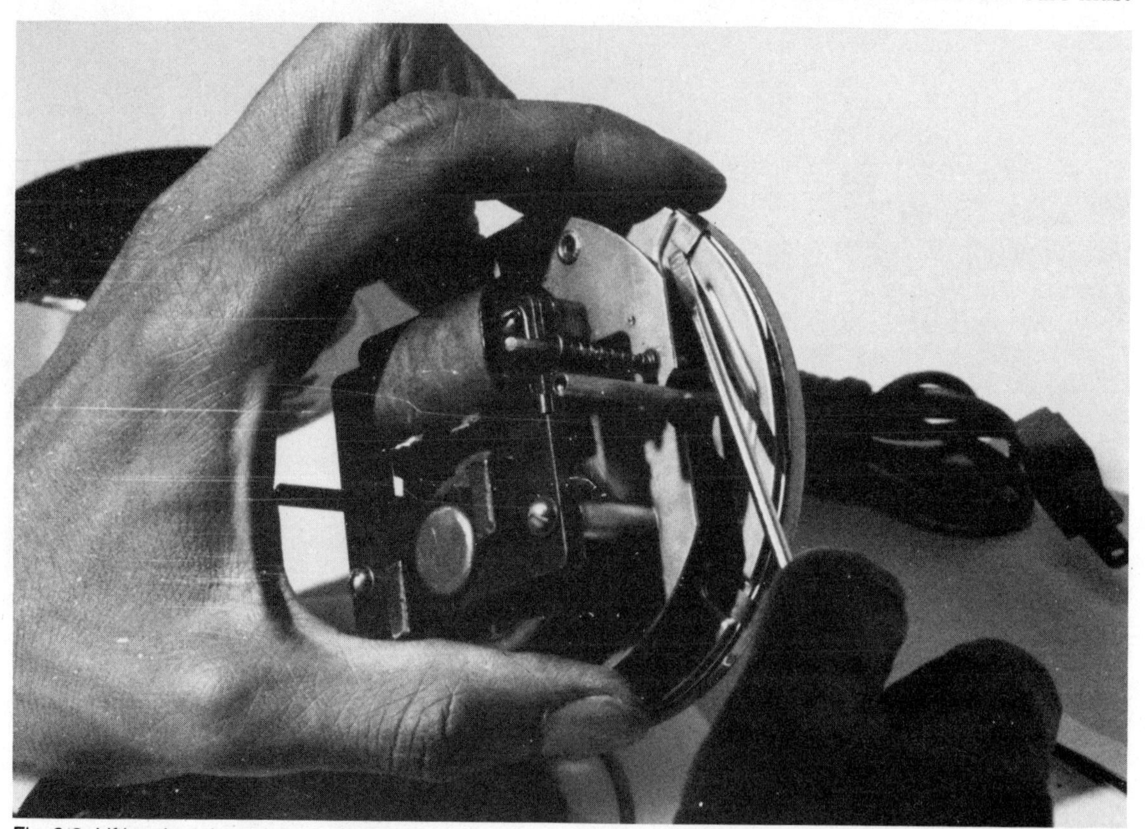

Fig. 8-2. Lifting the tabs to remove the bezel.

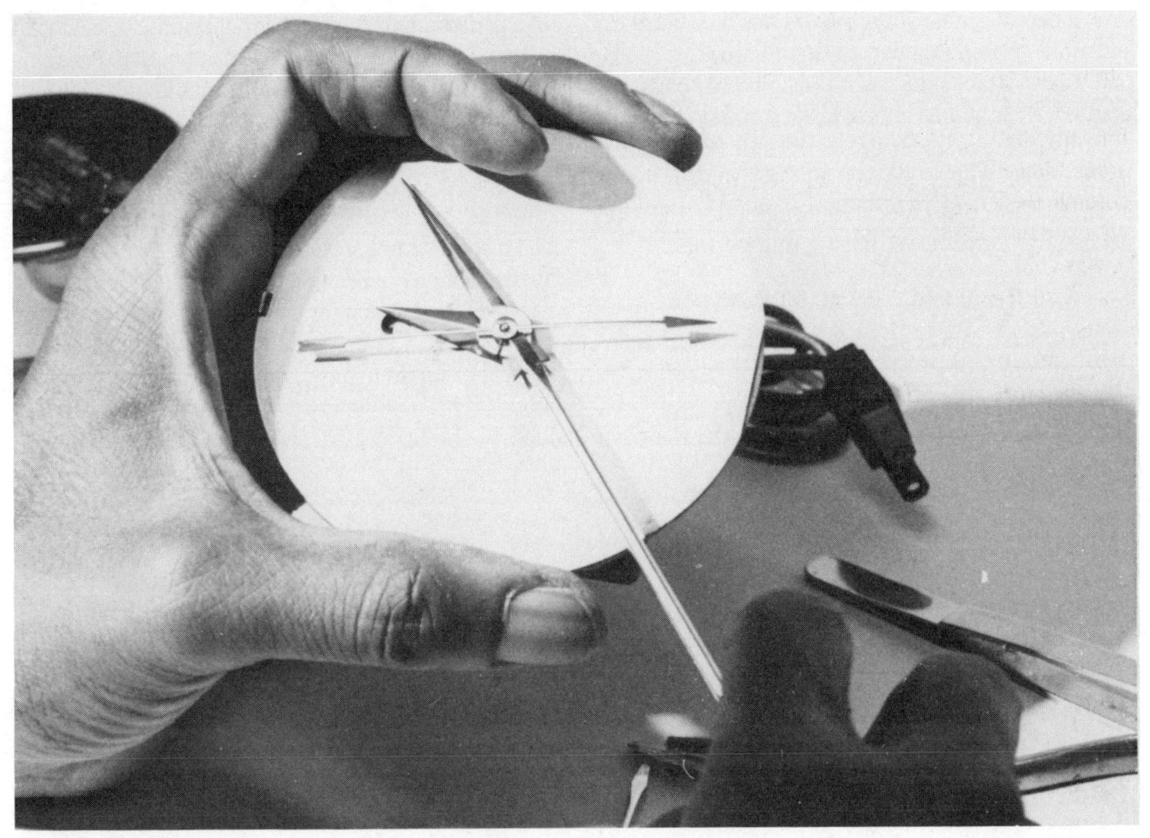

Fig. 8-3. Prying off hands to remove the blank dial face.

be exercised here not to scratch, bend, or break the hands. One clock repairer told me that he never used a hand removing tool because any inadvertent scratches just served to make the clock face look older! I recommend use of the tool whenever possible.

The front plate comes off with the removal of three screws revealing the going wheels. This was not an arduous task because little dirt and no varnish (hardened oil) had accumulated. Now was the time to remove and replace the motor and to fit the substitute arbor with its threaded end. At this juncture, the line cord had to be soldered to the leads sticking out from the coil. Any soldering mentioned within these pages has been done with a 25-watt, 120-volt, pencil-type soldering iron. Rosin-core solder is used rather than acid-core solder meant for heavier repair work (and that could harm electronic parts and circuitry).

Reassembly should go forward smoothly because you reverse the steps just cited. The top metal plate is positioned so that the hand-set arbor with its pinion fits in the proper hole. The three screws are put back, and the white dial face and the hands are replaced with the bezel and glass added finally. The movement goes back into the front of the marbel and holds the chapter ring when the cup is fitted in the rear.

There were no stains on the marble, but there was some grime that was cleaned with a little soap and water. Unless the clock is dropped or abused in some way, it should run at least another 30 years! The insulation on older line cords had a tendency to harden and crack, presenting the possibility of shorts. The cord, therefore, might be the only necessary replacement during this period.

TOOTH REPAIRS ON AC MOVEMENTS

"You can have this Sessions for $6," he said, "I can't make it run, and my wife says I should sell it."

I had dealt with the chap on numerous occasions prior to this encounter, so I picked the clock up and examined the exterior carefully. A beautiful specimen of a mantel clock with a familiar Sessions dial, and a fine wooden case in excellent condition, it seemed worth the price if it could be fixed. I bought it.

One of the inducements was that it was a "non self-start." A little lever had to be pushed down to spin a solid wheel and start the movement running. As mentioned previously, these movements are the electrics being sought after these days. Almost immediately, the movement came out of the case! The cord and its connection to the armature was in good condition. The next step was to plug it in and depress the little lever. There was movement for

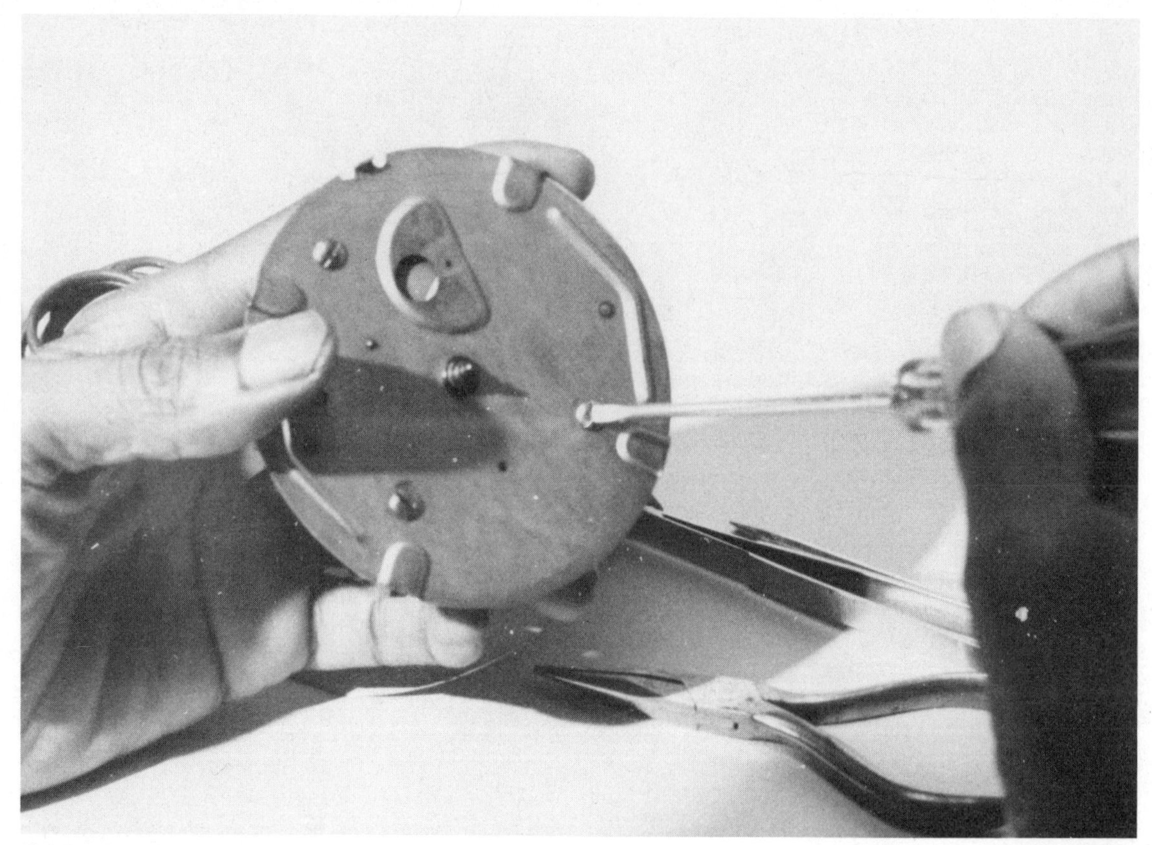

Fig. 8-4. These screws free the dial plate when removed.

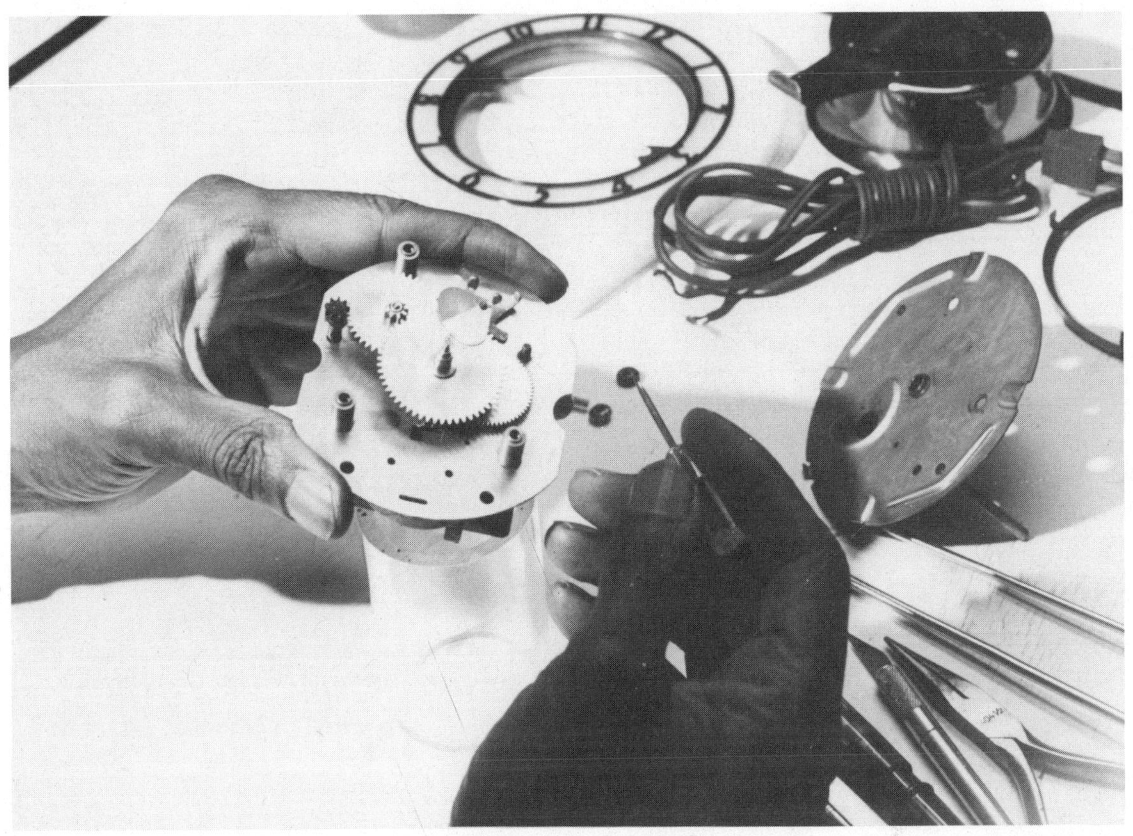

Fig. 8-5. A Telechron movement with top plate removed.

about a half minute and then everything stopped. A repetition of this action several times showed up the fault that had gone undetected by the previous owner. One of the wheels had three teeth missing! This is a defect that will hang up a movement the moment that blank area with the missing or damaged teeth comes into position where it should mesh with the pinion.

The accepted repair for this malady (other than replacing the wheel) is to file the stubs of the old teeth down, cutting into the wheel a little until there is a nice, clean gap between the teeth remaining on either side. Then, from a wheel with teeth of the same profile, a piece is cut that is soldered into the original wheel. Unless you have been doing this for years, the result will not be the tidiest job in appearance. Cutting the new piece with the necessary teeth is difficult. Here is another solution based on techniques clockmakers employed hundreds of years ago and are still recommended by some authorities today.

Back in the old days, whenever a pallet pin worked loose a length of needle the correct diameter was substituted. A dab of shellac was applied to the end of the new pin and it was driven into the hole in the anchor. It lasted indefinitely. Now here was a repair which, at first blush, did not seem practical. It was adhering metal to metal with a dab of shellac.

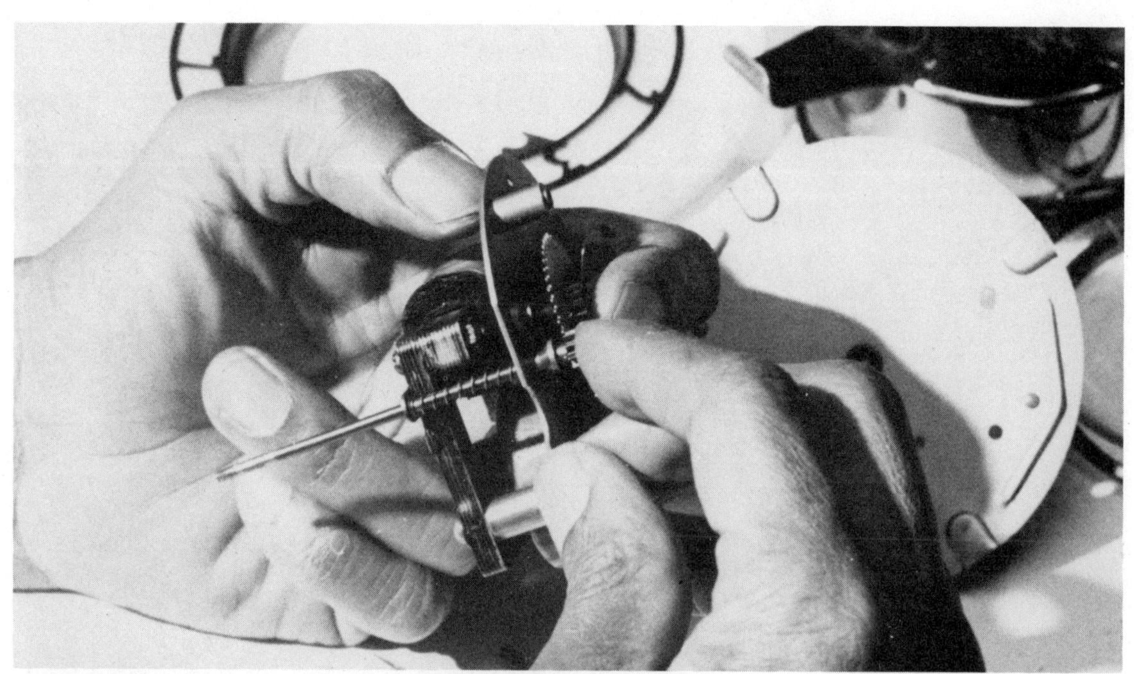

Fig. 8-6. A hand-set arbor must be inserted through the back plate, fitted with spring and then fed through holder on the motor.

Fig. 8-7. Pinion on a hand-set arbor (not visible) is repositioned in depression on a front plate.

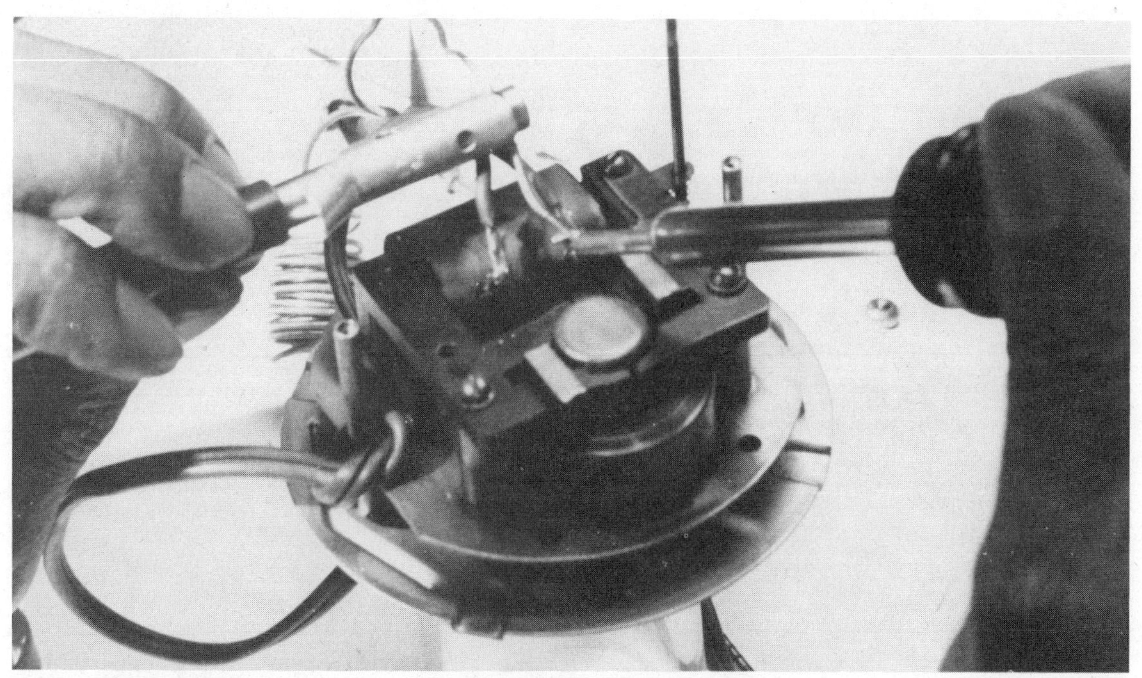

Fig. 8-8. Soldering the line cord on the leads from the coil. Solder is held by a little device called a Thumbsaver nail holder.

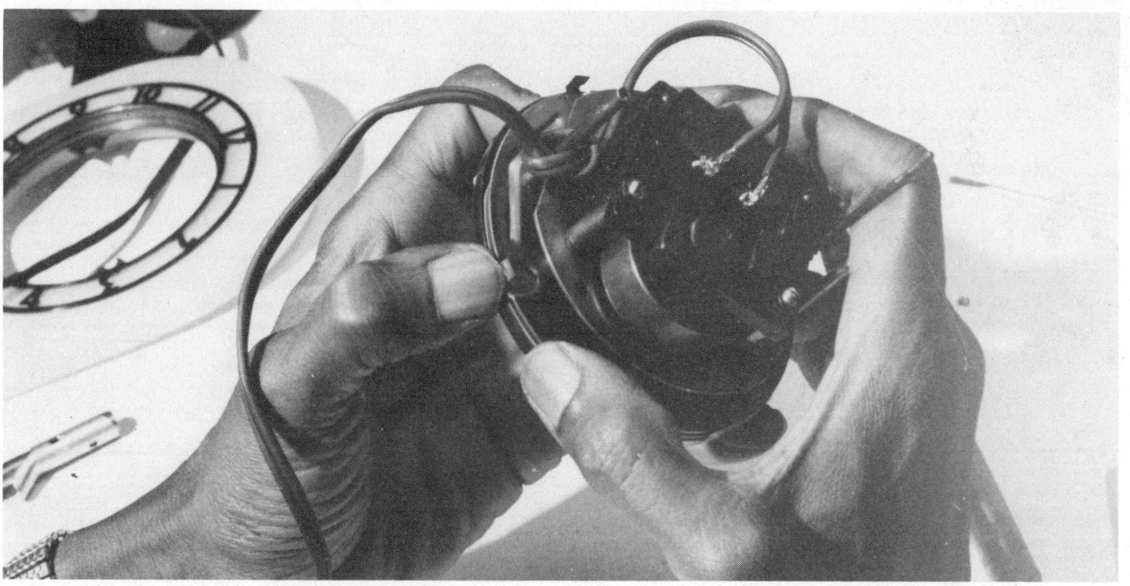

Fig. 8-9. Repositioning tabs for the bezel. Note the chapter ring with the retention ring inside to hold the movement.

Fig. 8-10. Non self-start Sessions movement. Note the start lever held by spring (arrow).

All right. If shellac would work, and it did, how about the new miracle adhesives—the epoxies and cyanoacrylate? They should be able to do the job a hundred times better. My findings are that they do the job and in a manner that leaves nothing to be desired.

Instead of cutting away at the damaged wheel, a new piece was laid right over the area of missing teeth. In this particular case the cyanoacrylate, Wonder Bond was used—just a tiny drop spread over one surface before applying to the wheel. Krazy Glue and other proprietary brands also can be used. Upon reassembly the clock ran, and it has run for more than three years.

Fig. 8-11. A wheel with replacement teeth cemented on.

Fig. 8-12. Replacement teeth on fiber wheel on a lantern clock.

First of all, if the pivot holes are not worn enough to produce excessive end play, the stress between the wheel and the pinion should be minimal. Figure 8-10 shows just such a repair on the Sessions (see also the section on the Lantern Clock in Chapter 7). In the case of the latter, the wheel (Fig. 8-11) was not brass but fiber. The piece used for the repair (patch) was taken from a plastic wheel from a discarded clock of recent manufacture. This also has been a permanent repair. Esthetically, it would have been better if another fiber wheel with matching teeth could have been found from which to glean the piece necessary for the repair. See Fig. 8-12.

LUX KEEBLER PENDULETTE

Little Pendulettes are very much in vogue today with some collectors. The escapement differs from that of most key wind movements. There is no anchor in the true sense of the word. The escape wheel, fitted with a lantern pinion, has V-shaped teeth spaced proportionately farther apart than is the case with most such wheels. The one-piece metal pendulum suspension rod rides on an arbor just in back of a tiny cup-shaped cylinder with a slot cut in it. As the pendulum swings to one side, a tooth drops into the slot and carries it to the end of the arc. When the tooth is released, another tooth hits the side of the cylinder even as the pendulum swings back. See Figs. 8-13 through 8-16.

Fig. 8-13. Lux Keebler Pendulette.

On the front end of the arbor (on this particular specimen) is a bushed-wire, hanger-shaped lever on the ends of which rode two kitten's heads in seesaw fashion. This little device goes through the face of the clock. The kitten's heads were missing on this Pendulette. There are parts supply houses that offer replacements made of paper.

Once the back cover was removed, it was immediately apparent that a thorough cleaning was in order. Disassembly is straight forward. The movement is fixed to the face with two screws visible beside the 8 and just below the 1. The minute hand is removed easily, but the hour hand is fitted on its arbor (the end of which is stamped or peened over the hand). This means that the hand, its arbor, and a wheel are lifted off the movement when the face or dial is removed. The hour hand, in this case, had worked loose so I allowed the wheel and arbor to remain with the movement for cleaning. The hanger arrangement slips off of its flattened pin away from the front of the clock, leaving the movement free to be administered to once the screws are out. The spring was let down (allowed to unwind) with the use of a bench key and the whole movement immersed in a bath of clock-cleaning solution.

Special attention had to be paid to the escape wheel and to the wheel that meshed with the lantern pinion because they had collected a hard coat of congealed oil. It took a good healthy brushing and another soaking to remove this accumulation. The pinions on all the other wheels required brushing.

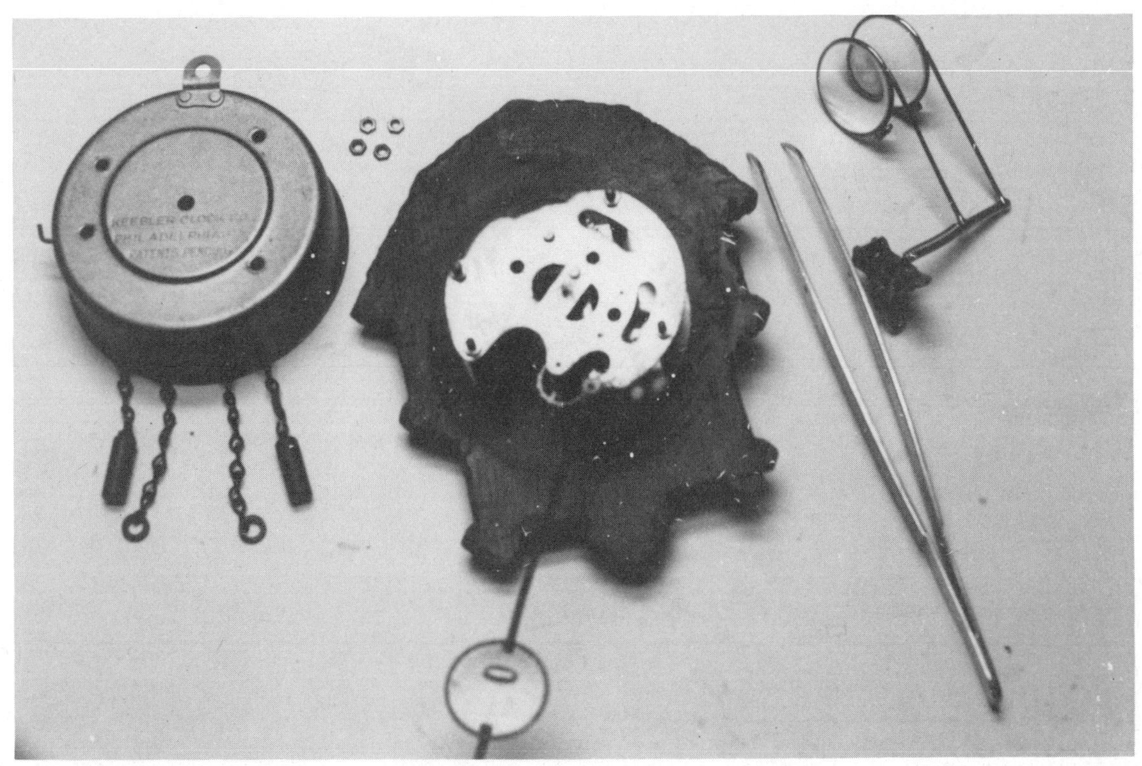

Fig. 8-14. The back of the Pendulette with the casing removed.

With the exception of the pendulum that was badly bent, no teeth were missing and no other parts were bent or broken. All that remained was to rinse the movement in warm water and to dry it thoroughly. Many books suggest that the movement be placed over a lighted electric bulb for drying, but a hair dryer emitting warm air is much faster.

As is often the case, the thorough cleaning and the straightening of the pendulum was all that was necessary to set the clock running once the spring was oiled lightly and a tiny drop of oil was added to all the pivot holes. Restoration to good working order was the sole endeavor in the case of this Pendulette. The face was cleaned gently with soap and warm water. The hands had been enameled originally and there was a bit of flaking off of this finish. No attempt was made to repaint them; gentle cleaning was considered adequate.

This particular Pendulette is not as highly prized as some of the others made by the Lux Company. It is, nevertheless, a worthy addition to one's collection and should appreciate as the years pass.

SCHATZ BATTERY-POWERED ROCKING CLOCK

A sorry, sorry sight was the Schatz Elexacta, shown in Figs. 8-17 through 8-26, when it came into my studio! There were copius quantities of rust behind the dial. Corrosion was rampant in the battery holder and the pendulum crutch bore no evidence of the original metal from which it had been

Fig. 8-15. Pendulette disassembled for cleaning and any necessary repairs.

fabricated because of a heavy coat of rust. The pendulum was missing as were the pillars and base upon which it had been mounted. Were it not for the fact that it was in one piece it could have been called a true basket case! Yet there were visible evidences of what a beautiful piece it was. The dial, almost elegant in its simplicity, was in excellent condition. The glass was unscratched and the brass bezel, with its ornamentation on top, could be revitalized with a careful cleaning. Even the hands were pristine. It was, indeed, worthy of salvage and total restoration once the necessary repairs had been made.

The rocking clock movement can be considered among those that are rewound electrically, in this case by an electromagnet. This one is a bit out

Fig. 8-16. The Pendulette rod and pendulum with a slotted cup at top, which is a vital part of escapement.

Fig. 8-17. Dial of the Schatz Battery Rocking Clock.

of the ordinary in that it is fitted with a pendulum rather than a pinwheel escapement. What makes it even more unusual is its ribbon-type, accordian-fold spring that activates a rocker-type lever to wind the spring when it runs down and closes the magnet contacts with a snap. The movement, electromagnet included, was encased in a plastic housing and, with the exception of the rust-covered crutch, seemed to have suffered no damage.

Applying current from a D-cell battery rigged with alligator clamps directly to the contact of the magnet and to a ground proved this to be true. The crutch oscillated back and forth rapidly. One end of the rocker moved down carrying a pawl with a contact on the end closer to a contact fitted to an arm attached to a metal disc on the magnet. The instant the contacts touched, the cap snapped toward the magnet shoving the pawl away (with the help of the ribbon spring), changing the attitude of the rocker and simultaneously winding the movement.

So, off came the plastic cover, and out came

Fig. 8-18. A view of the back of the Schatz Rocking Clock before cleaning and repair.

the battery leads, all green with oxidation. The hands were removed. The movement was then unscrewed and lifted out. Nothing was amiss with the wheels, the springs, or any components between the plates; the movement was set aside.

First, the battery case got a thorough cleaning to remove the corrosion that had leaked into it. Then the brass battery leads, which also serve as close-coupled contacts, were tackled. A Dremel Moto-Tool with a wire brush was used for this in conjunction with an X-Acto knife for scraping the hard-to-clean areas. When all was shipshape here

Fig. 8-19. The Schatz movement with the plastic cover lifted off.

Fig. 8-20. A preliminary check of the armature and movement.

Fig. 8-21. Lifting out the movement that separates from the electromagnet.

and the brass gleamed, the rusted crutch was tackled. It responded to my ministrations that included a thin coat of oil.

The back of the circular case upon which the dial and bezel is mounted will be cleaned thoroughly (all rust removed and spray painted brass as it was originally). The exterior rim that was painted a rich brown will be restored. Because it might not be possible to order a pendulum, one (similar to the one on the kit skeleton clock described in the kit section) will be made up.

The clock, it would appear, sat upon two columnar pillars, one in front of the other, rather than side by side as with the Anniversary clock. A brass base from a discarded Ansonia case plus columns of our own fabrication will be fitted. Research, thus far, has turned up but one parts supplier who offers replacement rocking clock mechanisms for sale with no pendulums being listed. Periodic maintenance will definitely be the order of the day with this lovely timepiece.

SHEFFIELD BATTERY-POWERED MOVEMENT FROM SPAIN

The intriguing features about this clock were its solid brass case with matching numerals and hands. It could be considered a golden oldie because the 4.5-volt battery was encased in a vaguely familiar red, blue, and white cardboard box with the brand name Eveready on it. A check of several large electrical supply houses confirmed the fact that this battery was no longer made. It just happens that three AA penlight batteries will kick out 4.5 volts so a substitution was made. That the clock was made by a company named Sheffield is an assumption engendered because this legend appears on the dial and is stamped on one of the plates in the movement. This movement, according to further markings stamped on it, is of Swiss manufacture. The words MADE IN SPAIN are also stamped on the plate.

A thorough cleanup of the original battery holder took priority. Lengths of wire had to be spliced to the leads from the movement so that an AA battery holder from a little portable radio could be installed. An elastic band was used to hold it in place, at least, temporarily.

No major repairs were required to the movement. It had a clear, plastic cup covering it so well that there was no evidence of rust or deterioration. The only work done to restore it to good running order was to adjust the contacts so that the circuit would be closed properly and effect proper winding of the mainspring.

There is one important measurement that must be checked here. It is the distance between the magnet coil and the armature (which should be about 3/64 of an inch). Adjustment is made by

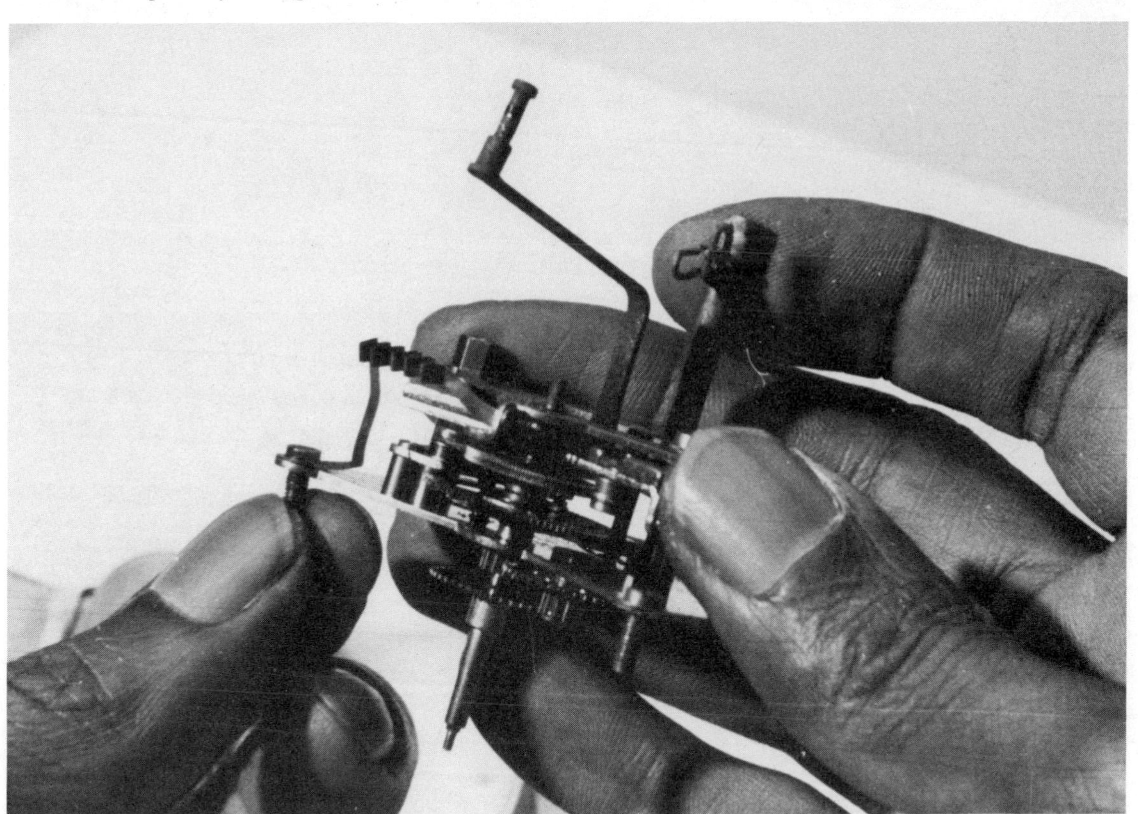

Fig. 8-22. The movement removed showing the ribbonlike spring and the rusted crutch.

Fig. 8-23. Sanding the corrosion off the edge of the battery contact.

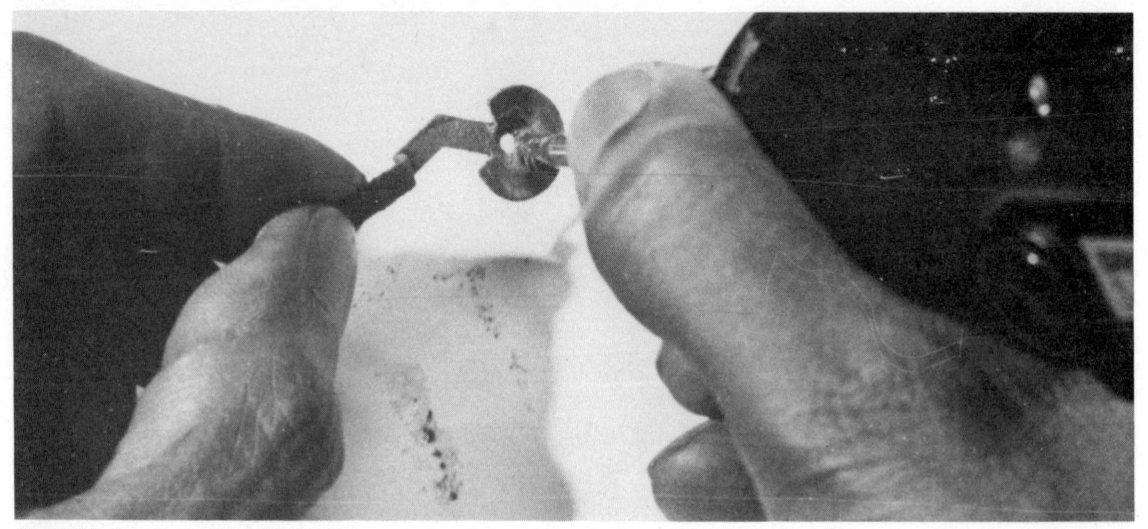

Fig. 8-24. Removal of corrosion using a wire brush in a rotary tool.

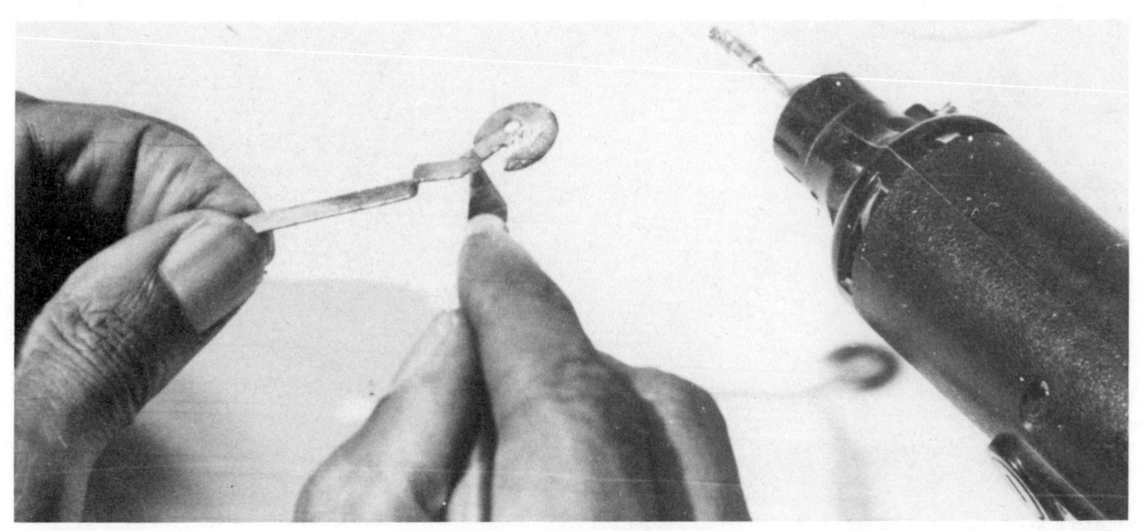

Fig. 8-25. Some of the corrosion required scraping with the back edge of an X-Acto knife blade.

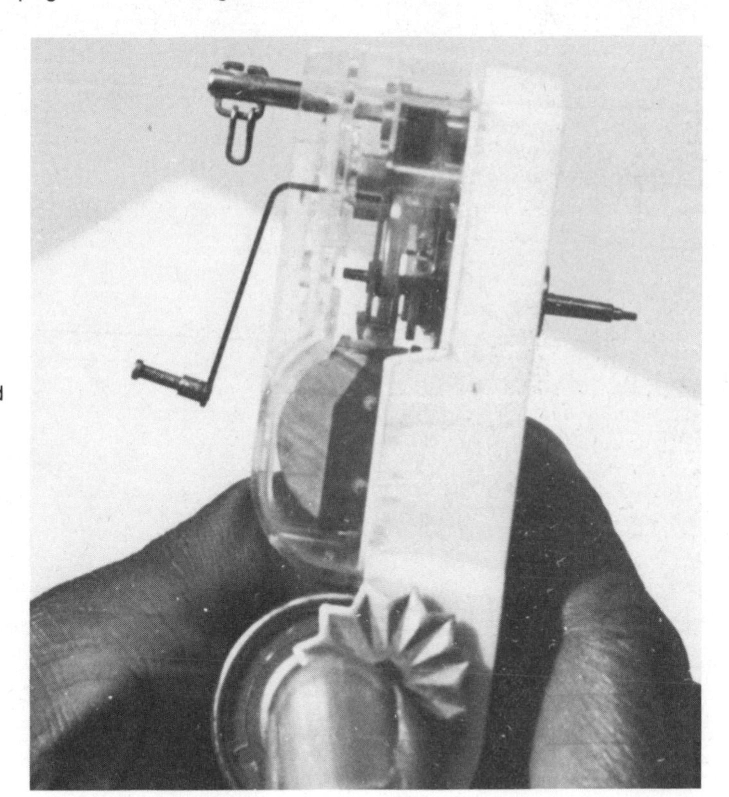

Fig. 8-26. The Schatz movement all cleaned up, adjusted, and ready for installation.

Fig. 8-27. The Sheffield battery movement from Spain, with a solid brass case.

Fig. 8-28. The back of the Sheffield movement.

Fig. 8-29. A close-up of a Sheffield movement showing adjustment screws for the armature and the position of the contact.

backing off on the two little screws on the guide plate and moving it back and forth. The movement (see Figs. 8-27 through 8-30) is quite similar to one made by Kienzle in Germany. The arrows shown in Fig. 8-29 point to the screws on the adjusting guide plate and to the contacts on the end of the armature assembly. The mainspring in its barrel is positioned just beneath the contacts.

Fig. 8-30. An extreme close-up of contacts and adjustment screws.

Chapter 9

Anniversary Clocks

THE POPULARITY OF THE 400-DAY (OR ANNIVERSARY) clock has escalated phenomenally over the past few years. The timepiece embodies all manner of paradox. Repairmen are always reluctant to tackle their repair and adjustment. Women are said to be totally enamored of their rotating pendulums and all the bright brass work. Yet the movement is one of the simplest one can encounter. This should be regarded as somewhat of an understatement in that problems do arise with the adjustment of the beat. The escapement is unlike any of the others discussed in this book.

I discovered three of these clocks for sale at a flea market. They were all piled together in a plastic box in abominable condition. There was but one brass base and it was mangled beyond repair. As shown in Fig. 9-1, the pendulums were off; the three suspension springs were broken or lost. Obviously, other parts were missing, but I could not ascertain that just by looking at the clocks quickly. It developed that the contents of that plastic box was available for $3. I bought the sorry mess, feeling that at least one whole clock would arise Phoenix-like out of the jumble in the box. See Figs. 9-2 through 9-13.

All other work came to a screeching halt! I delved and sorted and matched pieces for several hours. Everything visible had a layer of oily dirt but, surprisingly, none of the brass was deeply scratched. It seemed, eventually, that by cannabalizing I could assemble two clocks, although some parts would be needed.

I pause here to express a debt of gratitude to eminent authorities, Charles Terwilliger and Henry B. Fried. They are responsible for the book. *THE HOROLOVAR 400-DAY CLOCK REPAIR GUIDE*, that bailed this writer out whenever he got in over his head! Days of research in my personal library and in those within a 50-mile radius of my studio turned up nothing but the most cursory of

Fig. 9-1. Three basket-case anniversary clocks.

information, some of it erroneous. I restored to the most effective technique I know: the laying on of hands.

My good fortune lay in the fact that all three clocks were Kundo movements all evidently manufactured at about the same period of time, somewhere around 1949. I selected the one of the three with a square dial because this seemed to have the least wrong with it. A quick check of the mainsprings showed that none of them were broken; the one on this particular movement was let down and disassembly went forward.

The simplicity of these movements is born out by the accessibility of the click and the ratchet controlling the mainspring within its barrel. Once the spring was unwound, the bridge over the ratchet wheel was removed, as was the click and its retaining spring. The next few procedures, it will be noted, are much the same as for most other spring-wound movements. The arbor that carries the hands is drilled for a pin that must be removed to free the dial. There are three pillars on the back of the dial that also hold it to the front plate with pins.

Separating the plates requires the removal of four pins, one on the back end of each pillar. Screws are fitted to recessed, threaded holes in the front end of each of the pillars. Once the pins are removed, the back plate can be lifted off. There was

Fig. 9-2. The front of a 400-day clock movement.

Fig. 9-3. A side view, showing wheel placement, on a 400-day clock.

Fig. 9-4. A back view of a 400-day clock with the proper bench key to let the mainspring down.

Fig. 9-5. Movement with the top plate removed showing wheel placement.

all manner of crud, gook and varnish on everything, of course, so the barrel housing the mainspring was lifted out and the back plate was replaced. The wheels were fitted into their pivot holes once again and the movement, without the mainspring, was lowered into a bath of clock cleaning fluid.

The cleaning fluid has a pronounced, acrid ammonia smell and should not be used unless the work area has excellent ventilation. By using a vessel deep enough to accept and cover the whole movement without any part sticking above the rim, a thick piece of cardboard can be placed over the bowl for the 15-minute soaking period.

All of the other brass parts, the pendulum, the platform between the movement, and the support pillars got a good healthy soak at this time. The movement was lifted out during the bath period and the components were brushed with a toothbrush dipped frequently in the bath solution. Out came the movement for a rinse in mild soap and water. It was set on paper towels and a stream of warm air from a hair dryer was directed at it. All of the brass gleamed at this juncture, but the steel arbors, the anchor pin, and the pallet all showed signs of rust. This required removal of the back plate once again to deal with this rust removal from each piece individually. While everything was "down," the two plates were polished and the pivot holes were

Fig. 9-6. An escape wheel and anchor with adjustable pallets and anchor pin.

cleaned. The movement now was a joy to behold, but the work had just begun. The remaining parts were cleaned and polished and it was time to tackle the escapement.

The pendulum is not at all conventional. It usually consists of four balls on curved L-shaped hooks all fixed to a brass staff from which they can be pushed outward with a knurled disc. The assembly hangs on a wire and rotates slowly. It is a very pleasing and even soothing movement. It is called a torsion pendulum. Instead of the balls, some clocks have a thick disc suspended much like a chandelier that also rotates instead of swinging from side to side. The problems with these clocks are traceable to the top hanger for the pendulum wire that is called the beat adjuster block. It shows clearly in several illustrations in this chapter. Within it is a weird looking piece of brass fitted with two tiny screws. The suspension wire goes up between these screws which, when tightened, clamp it firmly. This piece, in turn, is hung on a screw between the arms of the block that can be moved to the left or right. Moving it either way adjusts the beat; this little operation takes a lot of time and patience.

There is a block at the bottom of the wire, too, and this is where problems arose for me because two of the bottom blocks were missing. While just about any part for this clock can be ordered from the Horolovar Co., Bronxville, New York or from one or two other materials suppliers, I chose to make one up from scratch. It's quite a trick if you can do it.

Select a piece of brass rod no greater than 5/64 of an inch in diameter and 9/32 of an inch long. Drill a hole about 2/64 of an inch in diameter through the piece equidistant from either end. Now turn the piece so the hole is parallel with the work table and drill another hole, centered, right down until it opens into the first drilling. Do not go all the way through. The second hole is threaded and fitted with a little flatheaded screw. The end of the suspension spring is fed into the original hole and the screw is tightened down upon it. It works! The pendulum can be hung with minor adjustments so that it is straight.

The suspension wire has to hang straight with no kinks or bends in it. And it must be a certain thickness that varies according to the clock to which it is fitted. This measurement ranges from .0023 of an inch to .004 of an inch. Use a micrometer for this. These suspension springs are affected by temperature changes, but the Horolovar Company offers springs that compensate for this. I miked a number of springs from good quality discarded watches and found two between .0032 of an inch and .004 of an inch and used them.

The next crucial step is to refit the fork that is

screwed to the suspension spring near the top. It controls the pallet pin that is about 1½ inches long. The fork must be a certain distance from the top block or a condition called *fluttering* occurs.

Once the clock is all assembled, the beat can be adjusted by turning the top block a hair to the right or left. There are two peep holes in the back plate through which the escape wheel and the anchor can be observed. The pendulum is given a gentle push to cause it to turn, and then the anchor is watched through the peep holes. The pallets should drop the same distance on either side of the escape wheel as they pick up and drop teeth. When the pallet on one side digs in deeper in the space between the teeth than the other, the movement is said to be out of beat and the adjustment of the top block mentioned above is called for. If several teeth suddenly skip by the anchor, the flutter condition

Fig. 9-7. A beat adjuster block with a fork clamped to a suspension spring.

Fig. 9-8. One unit completely assembled with the pendulum positioned.

exists and is eliminated by raising the fork on the suspension spring a hair. These are the time-consuming adjustments that require great patience. Sometimes that flutter really gets going and the hands can be observed moving around the dial at such a pace that the timepiece gains an hour in 20 minutes!

The pallet pins are attached to the anchor with two screws for each pin. They are adjustable on most clocks, but should be moved only as a last resort. Both clocks discussed here were set up and made to run accurately without moving the pallet pins at all. The movement reassembles quite readily. Most of the time expended on the clock will be devoted to beat adjustment. The following are some hints and clues picked up in making two working clocks out of three basket cases.

□ Timekeeping is regulated by turning the knurled knob on the pendulum to extend or retract the four balls. Turn them in if the clock runs too slowly, or turn them out if it runs fast. If they go to their extremes in either direction with no change in the accuracy of timekeeping, a suspension spring of different thickness must be fitted. Use a weaker spring to make it go slower or a stronger spring to speed it up.

□ The swing of the pendulum is usually measured by just how far it turns in one direction in degrees. Here no one seems to agree. Some authorities say it should turn a full 360 degrees. Others suggest just 270 degrees minimum, but as much as 450 degrees maximum. Somewhere else it was suggested that the pendulum should make just eight complete oscillations per minute. Both of the clocks renovated here came amazingly close to the eight-oscillation figure.

□ The mainspring is enclosed in a barrel with a removable top. One top was missing so that spring was not used. The covers were taken off the other two, and inspection showed that no rust and very little dust or dirt had accrued. Removing the spring

Fig. 9-9. A dial with brass chapter ring and cutout figures all in one thin brass stamping (courtesy of Mrs. Gerri Miller).

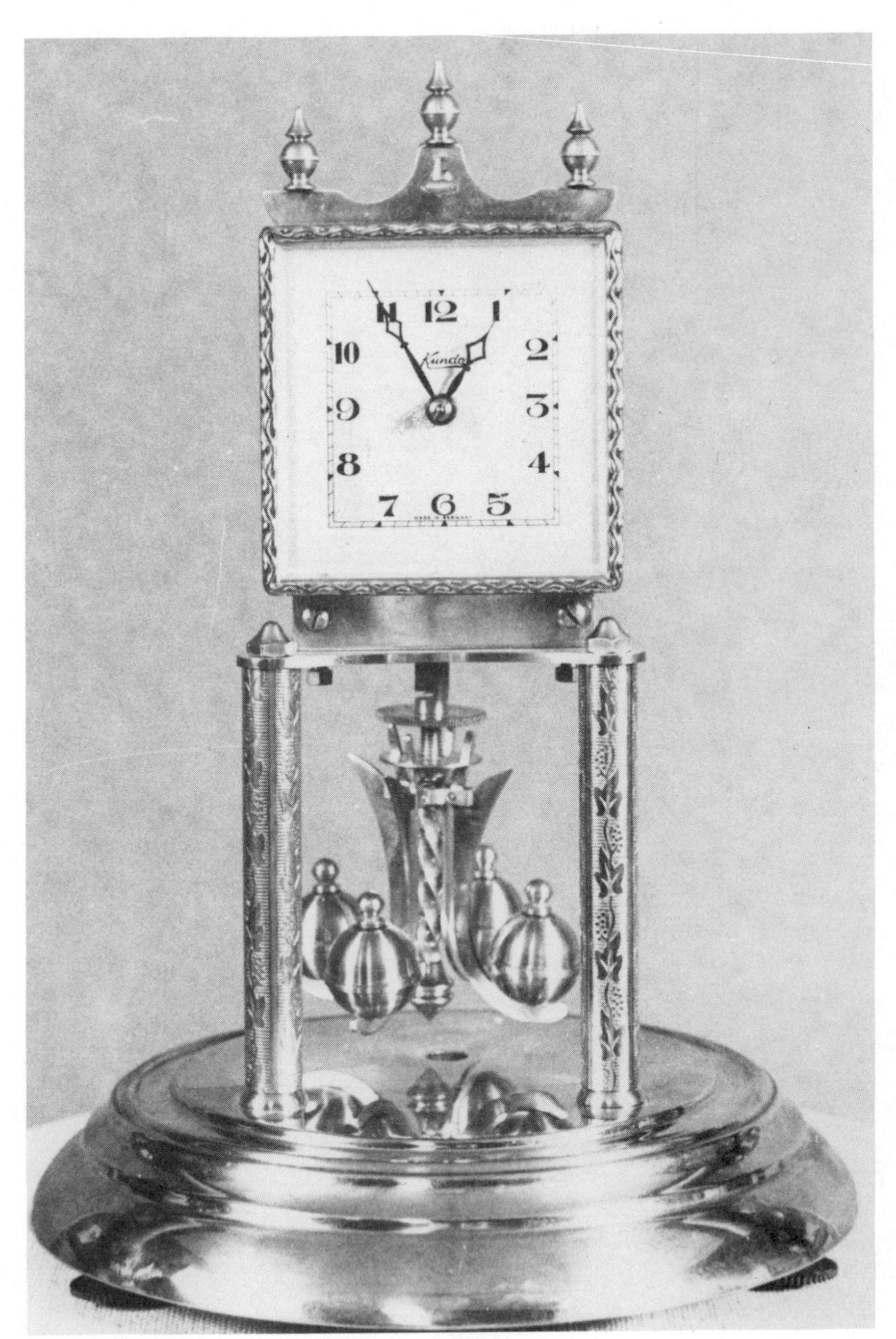

Fig. 9-10. One of the basket cases completely restored.

Fig. 9-11. Another of the repaired and restored 400-day clocks. The wooden base was made on a lathe.

Fig. 9-12. Discarded wooden disc from which the 400-day clock base was lathe-turned to size.

from the barrel and then replacing it is quite a chore without a mainspring winder. If the possibility of touching more than one or two anniversary clocks in a lifetime is pretty remote, expenditure for the tool seems unnecessary. If the spring is broken, a new one can be ordered that is coiled to so small a diameter that it is said to slip easily into the barrel. For me, the occasion has never arisen for this type of spring or a mainspring for any other clock to be wound by hand.

☐ The old adage about the use of as little oil as possible applies here. The pivot holes get a tiny drop as do the pallets and possibly the mainspring. Anywhere else is excessive. All the wheels are exposed between the plates and oil will surely pick up dust very easily under these circumstances.

☐ There were no usable brass bases with the clocks I purchased, but I found one that was a perfect fit for one clock. Using it as a guide, I turned up another on the lathe out of some beautiful wood just so the clock could be displayed until a proper base was located.

Fig. 9-13. A pair of beauties from the scrap heap! Both keep excellent time.

In the center of most bases is a little brass cup. This is raised to keep the pendulum from bobbing about when the clock is moved. The tip of the pendulum should not sit down in it.

□ If, in adjusting the beat, everything seems to work to no avail, it has been suggested that the anchor pin be bent slightly toward the back plate. This should be tried only as a last resort because it can cause other troubles without curing the out-of-beat problem. Bending it could simultaneously tilt it to one side or the other and returning it to its original position could result in its breaking off completely.

□ These clocks must run in a perfectly upright position. The bases generally have three adjustment screws built into them just for this purpose. Anything substituted for the original base must be set perfectly level even for the periods when the clock is being repaired and checked. I recommend use of one of those little round levels (about 1¼ inches in diameter). It will come in handy for work with other clocks as well.

These timepieces are such a joy to observe that some owners cease to be concerned about the accuracy of their timekeeping. Since World War II, anniversary clocks have been brought into this country from Germany in large numbers. Several American manufacturers make them for sale now and, in essence, they are not to be considered collectibles. Nevertheless, they are fairly expensive and they are worthy of repair and restoration.

Chapter 10

Cuckoo Clock Repairs

FIRST OF ALL, LET IT BE KNOWN THAT CUCKoo clocks are quite like many other spring-wound timepieces. Exceptions are that they are fitted with a pair of bellows in the back and they have little figures or birds that pop out of doors above the dial in front. Some have a music box hidden in the very top of the case. All of these devices have wire hooks and pins and cams to activate them. Once the movement is removed from the wooden case, it can look most intimidating. Our firm advice here is not to separate the two plates! Returning everything to its proper place might prove well nigh impossible for the layman.

Here again there is a dearth of written material on the subject. Most clock repair books devote no more than a page or two to cuckoo clocks. There is a humorous little self-published pamphlet entitled *How To Train A Cuckoo* by Malcolm C. Gerschler. It is available by writing to The Wag On The Wall, 2005 Valle Vista, National City, CA 92050. Several cuckoos were set up in my studio before I came across this pamphlet, but it still proved helpful.

The first specimen encountered came in because someone had turned the hands backward and, in so doing, had forced one of the chains off its pulley. This was corrected by thrusting and manipulating tweezers through the inspection hole in the side of the case, but this accomplishment was more good fortune than skill.

The second one had real troubles! The weights could not be lifted. The pulleys seemed to be jammed. It was decided to take the movement out of the case and to perform whatever surgery was deemed necessary. The procedure for this follows.

The clock shown in Figs. 10-1 through 10-14 had both the bird and the little man behind doors up at the top. The bird is tied to the operation of the bellows while the seated man does his thing in conformity with the striking segment of the movement. There is a third weight on this movement that

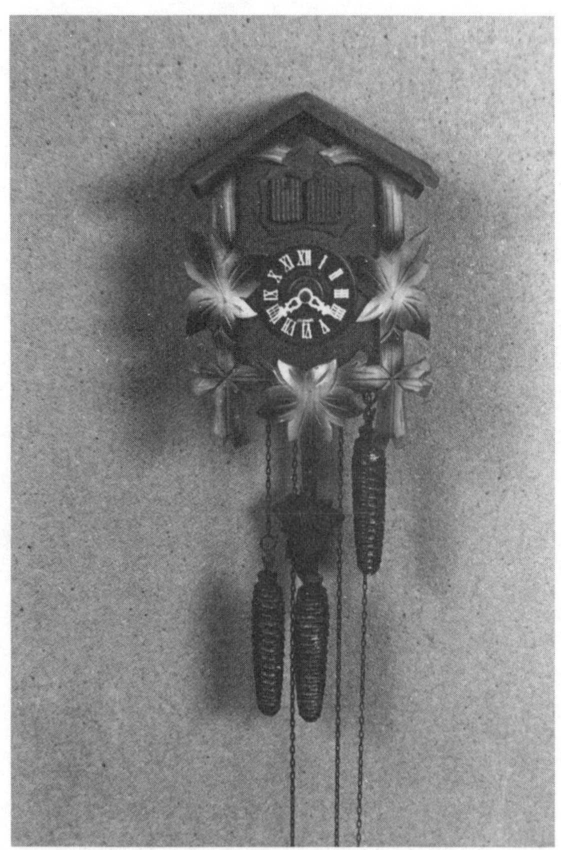

Fig. 10-1. This three-weight cuckoo clock was totally inoperable (clock courtesy of Loretta Holz).

powers the music box. Both the bird and the little figure come forward through the use of levers and, in so doing, push the doors open with the aid of little wires attached below them and fixed to the doors. If the adjustment is off by a little bit and the doors do not close all the way, the tendency is to push the door closed. That can mess up a lot of things inside.

The wires have to come off the doors. The hands must be removed; a knurled nut holds the minute hand on while the hour hand is press-fitted. The weights are removed, and the clock is turned around for removal of the wooden back plate to which the gong is attached. It is apparent immediately that there is no access to the movement unless the bellows comes out first. The wire connections are removed from the hammer tails that come from the pinwheel. The bellows are mounted against the upright outside walls with a nail and a wood screw. Once they are removed, it is just a matter of reaching in past the movement to the brackets with which it is screwed to the front of the case.

The chains can now be removed in preparation for the next step. The trouble with this clock was apparent immediately. Someone had oiled it liberally and laid the groundwork for an unbelievable accumulation of varnish (congealed oil mixed with dirt). This movement got a clock-cleaning solution bath and soak the likes of which few clocks ever get. The bird and the figure came off the levers first. Note that nothing has been mentioned about tampering with the strike mechanism on the front plate! The strike mechanism is of the rack-and-snail variety (at least on the modern versions) and if nothing is moved the hands can be set to match the striking without too much difficulty.

The varnish lay on the plates, between the leaves of the pinions, in the chain pulleys and just about everywhere. It required brushing and probing with toothpicks to remove. In an attempt to overcome the resistance encountered in raising the weights, someone had bent the click mechanism on one pulley. Rather than endeavor to separate the plates to remove the pulley, adjustments were made with a screwdriver and a parrot-nosed pliers. All of this was slow and tedious, but moving the pulley wheels after the movement was dry activated both trains and the fly whirled merrily. Reassembly was in order.

Here, again, putting the movement back in the case and adding the bellow and hooks, etc., is a reversal of removing it. The levers for the bellows are of slightly different length and must be placed correctly so that the low and high notes of the cuckoo are not reversed. But before the movement

Fig. 10-2. The wire linking the bird to the door is removed.

Fig. 10-3. A knurled nut is removed and the minute hand is lifted off.

went back in, the music box got some much needed attention. The arbor from the drum with the music needles was bent and this caused the main pulley to bind against the case. When the arbor was straightened, the pulley was replaced and revolved without any binding. The comb, too, had to be adjusted forward slightly because some of the pins on the drum were worn. If moving the comb forward just a hair does not make all the notes sound, the music box mechanism should be replaced.

Once the movement is properly mounted and the bellows have been replaced, everything from here on is a matter of fine adjustment. The wires that tickle the bird and the figure into performing must be carefully bent back into the proper position or they will hang up either the bird or the bellows. The chains must be fed through the holes in the bottom of the case after replacing them on the pulleys. The wires for the doors must be refitted. The wood on the doors is very soft so care must be exercised so as not to pull the hooks out. Now the hands must be refitted.

This operation is not all that difficult. If nothing is askew in the striking mechanism, replace the hands and hang the clock. Now turn the minute hand until an hour strikes and then just set the hands to read that hour. Start the pendulum going and turn the hands to the correct time. There is just one

Fig. 10-4. The hour hand (press fit) is removed.

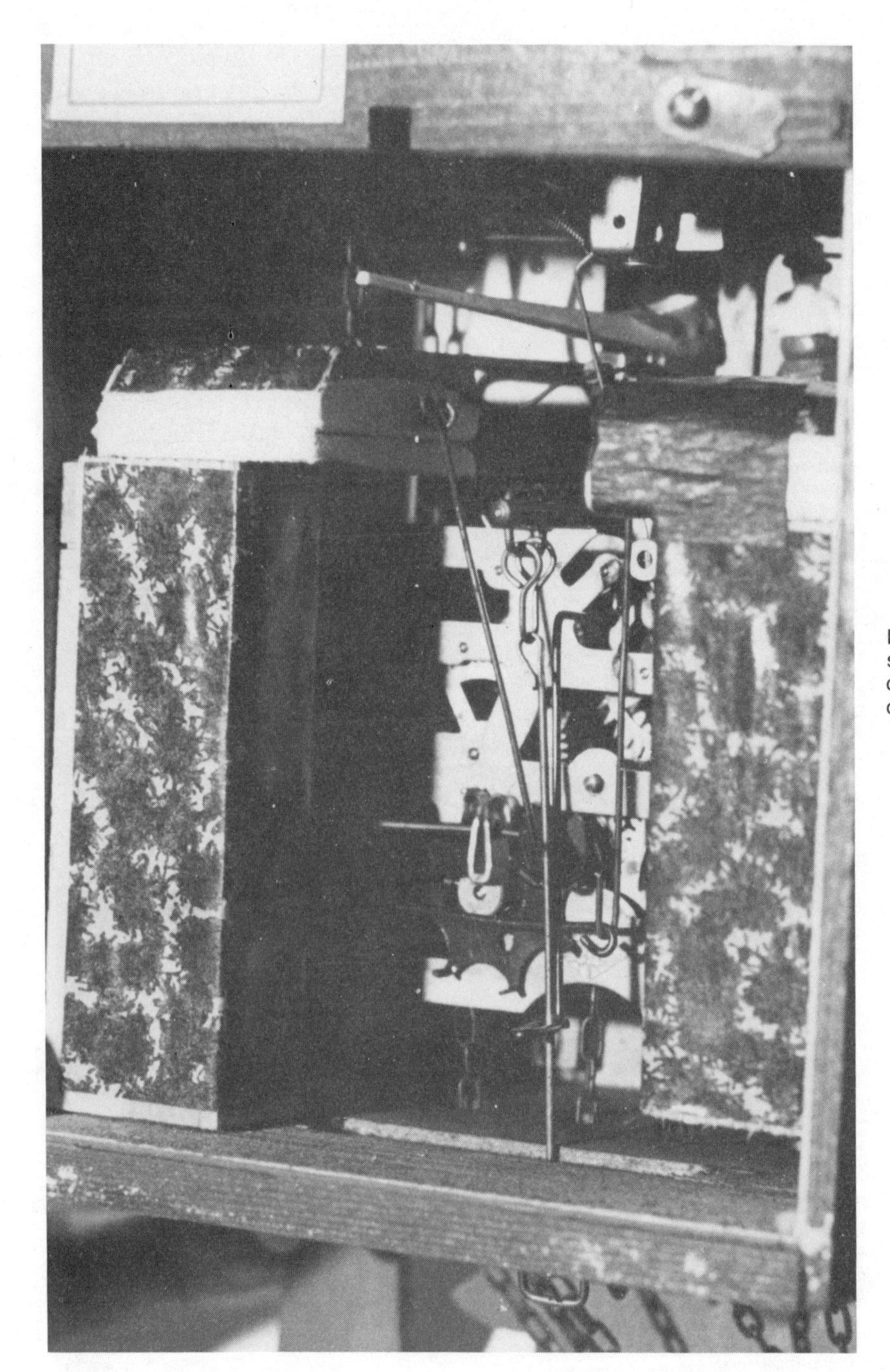

Fig. 10-5. The movement is screwed to the front of the case so bellows has to come out first.

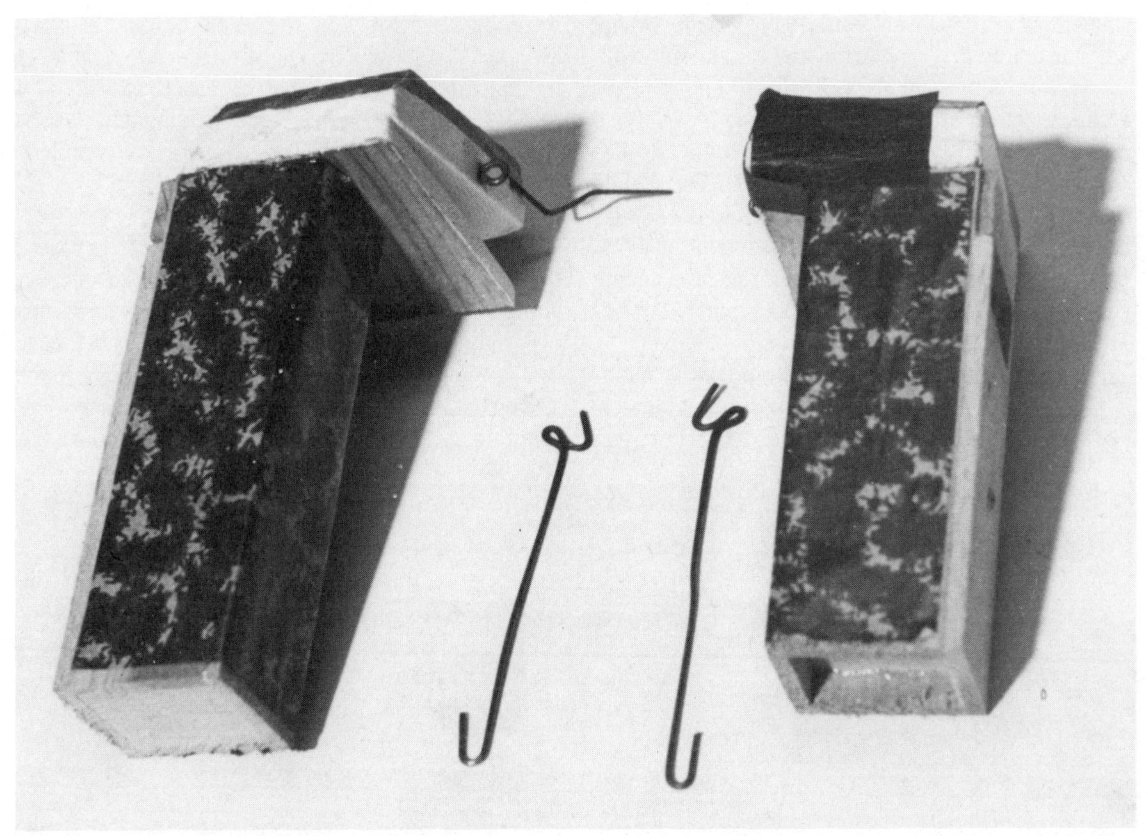

Fig. 10-6. Bellows, pipes, and lifting wires after removal.

thing that should be clarified about setting the hands. The minute hand has a brass insert. If the knurled nut is tightened properly, it might be difficult to get the strike to occur exactly on the hour. I got it down to about 2 minutes on either side of the hour and had just about given up when I came across Mr. Gerschler's little pamphlet. According to him, the minute hand has to be removed and the brass insert (if there is one) repositioned so that the hand's relationship to the striking is changed for the better. This takes a bit of trial and error but it can be done. If the insert cannot be moved without the possibility of destroying the hand, discretion becomes the better part of valor. In other words, leave it alone. This remedy worked in the case of the clock worked on here.

Once all systems are go, just hanging the clock and swinging the pendulum a bit is not enough. The pendulum must swing the same amount to either side or the clock is said to be "out of beat." This can be done by moving the bottom of the case to the right or the left just a hair at a time until the tick sounds just right and the pendulum does not slow down or stop. The sound of the "tick" should be just as robust as the sound of the "tock." The rate of running (fast or slow) is regulated by moving the leaf on the pendulum up (faster) or down (slower). Determine that the clamp that holds the leaf to the rod is tight or the normal vibration of the movement will cause it to drop and throw the timekeeping off.

Everything went so well with this clock I decided to clean up another that had been lying around. The procedure from start to finish was exactly the same except that when I put the weight on the going chain it went right down to the floor! I removed the wooden back plate and went probing. Imagine my surprise to discover that the anchor that was spot welded beneath its arbor had fallen off!

Now here was a dilemma. Would soldering be a good substitute? Fortunately, all my previous experience with clock repairs indicated that there was little or no stress on the anchor. Why not try cyanoacrylate (which is advertised as capable of holding 5000 pounds)? So the arbor was filed to remove traces of the previous welding and the anchor was also cleaned. The adhesive was applied, the two pieces were fitted together, and then allowed to dry.

Here again there was no inclination to separate the plates to get the particular arbor out. Inspection showed, however, that the back plate has a tab into which the pivot for this particular arbor is fitted. It became just a matter of bending the tab back enough to free the arbor and the repair could be made. The arbor was replaced and this time, instead of the

Fig. 10-7. The music box is housed in the top of the case. Positioning in this housing (used just for photography) differs from that of the specimen worked on.

Fig. 10-8. The movement with chains comes out after removal of four little wood screws.

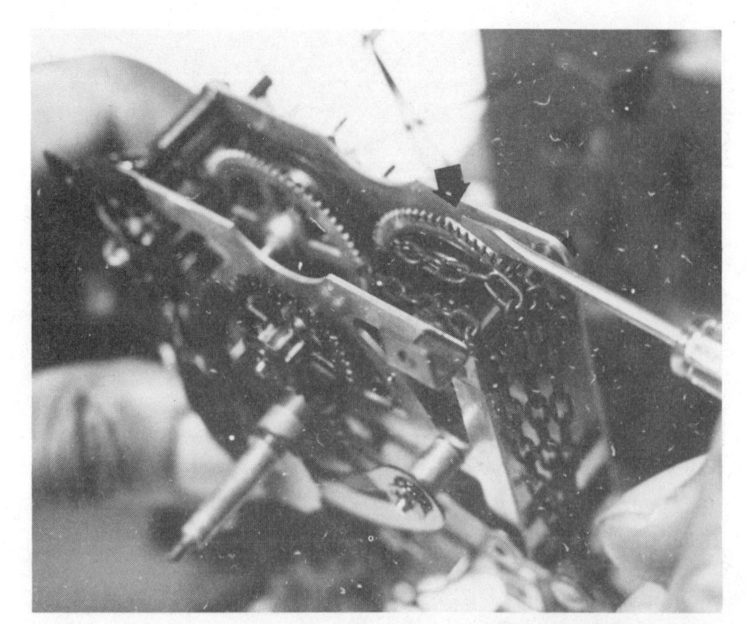

Fig. 10-9. The click had to be repositioned to keep weight from falling each time it was raised.

Fig. 10-10. The front view of a cuckoo movement.

Fig. 10-11. The rear view of a cuckoo movement.

Fig. 10-12. One method of supporting a movement for oiling or other work.

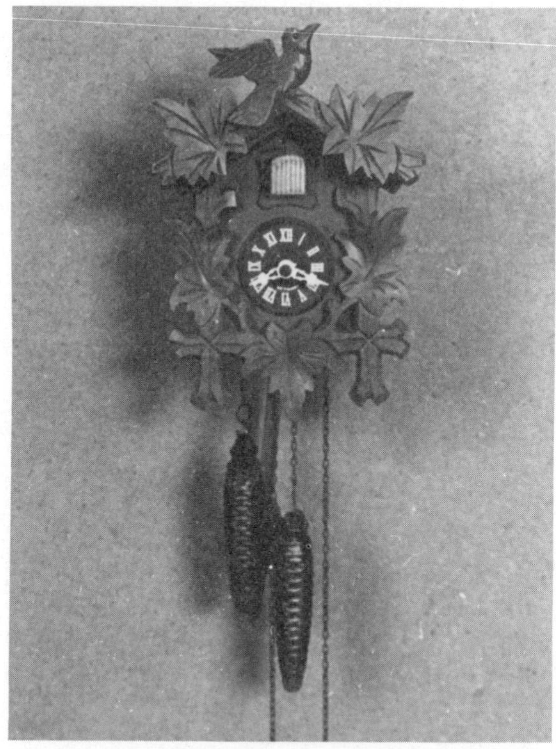

Fig. 10-13. A two-weight cuckoo clock has no music box and a single door for the bird over the dial (clock courtesy of Don Chaiken).

weight falling, the clock ran as it should. It has been running for more than six months now. The ironic aspect of this is that the anchor on many German alarm clocks is amazingly similar and one could well have been modified to fit.

Fig. 10-14. Two- and three-weight cuckoo clocks after cleaning, repair, and reassembly.

Parts such as this one can be ordered from some materials and parts suppliers. They offer a wealth of weights and birds and sundry other parts as well.

There is still another working cuckoo movement lying around the studio here. The weights are missing and the bellows are sadly in need of repair. It might be an excellent exercise to encase it in acrylic and use plastic cylinders filled with BB shot for the weights. It will be an interesting project if time ever permits.

Chapter 11

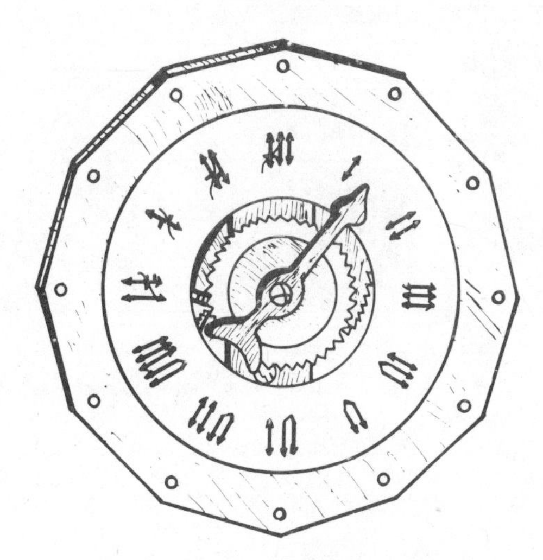

Alarm Clocks

THE ALARM CLOCK IS ABOUT AS AMERICAN as apple pie! During the 1920s and the 1930s, the Sunday comics in the newspapers were full of characters who threw alarm clocks out of windows or bounced them off the heads of other characters. Humor? They were, and frequently still are, so inexpensive that repairers are very reluctant to work on them. Now the older ones have begun to be considered as antiques or, at least, collectibles.

Quite like most other clocks, repairs can be made on an alarm clock (Figs. 11-1 and 11-2) once the movement is exposed. Start with the removal of the winders (the keys for time and alarm). There are arrows usually indicating which way they should be turned to wind the clock. Turning them in the opposite direction takes them off. On occasion, these keys will have to be pried off. This is done with the aid of parrot-nosed or similar pliers. The next step differs with the make of clock. A quick inspection will determine the way to go. Most often the bezel has to be pried off (not always that easy) and three or four screws must be removed from the back of the case. The movement should come right out at this juncture to be examined.

Just such a clock was brought into my studio recently. It had run for more than 20 years before being dropped on the floor suddenly. The movement did not even have to be removed from the case to show that the balance wheel with the hairspring had been jarred loose.

The fix here is the same as the next step in dismantling one of these clocks. One of the screws (a threaded steel cup) is backed off several turns; this normally releases the balance wheel. See the arrow in Fig. 11-3. The end of the spring is held in a stud with a brass wedge that has to be removed. A kink or bend is usually left in the end of the spring, and this should be used as a guide for proper replacement. More accuracy can be obtained by putting a scratch on the end of the spring at this point

Fig. 11-1. A variety of older alarm clocks made by (from left to right) Westclox, Lux, and Phinney-Walker (Germany).

Fig. 11-2. A double-bell, key-wound table alarm clock now being manufactured and sold by Timex Clock Company, Waterbury, Connecticut.

Fig. 11-3. A West German alarm restored, but with the dial removed to show brass plates and wheels. The movement is fitted with a seconds hand (arrow).

Fig. 11-4. A Westclox movement with the bell mounted inside housing.

with a tiny screwdriver or a sharp, pointed instrument.

In this case, retracting the screw provided space to refit the wheel. Before replacing it, the pivots of the arbor or balance staff were checked. They are supposed to feel sharp to the touch. If wear is indicated, they should be sharpened or ground to a point. Replacement of the wheel in this case started the clock running immediately. See Figs. 11-4 through 11-10.

If inspection shows nothing obviously wrong visibly with the movement, the next step is to clean it. Just place the whole movement, without the balance, of course, in a container of clock-cleaning solution deep enough to cover it completely. Scrub it a while with a cleaning brush while paying special attention to the periphery of the escape wheel and all the pivot holes. This bath and cleaning should always be of about 15 minutes duration.

Remove the movement, subject it to a rinse in mild detergent soap and warm water, and then dry with the warm air from a hair dryer. Now do the same thing to the balance wheel and spring. Inspect the movement carefully to make certain it is completely dry and set the balance into place. This calls for a little manual dexterity the first few times around. The balance is held upright in the cups with tweezers and the top one is screwed down until the balance exhibits a minimum of end shake (movement from side to side). The end of the spring is replaced in the stud and held there with the brass wedge.

At this point, the movement should be oiled carefully. Both mainsprings (time and alarm) should be wound fully and the closed coils should be covered lightly with oil. All the pivot holes require attention next, and then alternate teeth on the escape wheel plus all of the teeth on the alarm wheel. If all is well with the movement, the balance wheel should start its oscillation and keep moving.

Continue the inspection. Take a good look at the pallet pins to make sure they are straight and not worn. If worn they can be driven out and replaced with short lengths of steel needles. It should be remembered, however, that this is an inexpensive clock that is being repaired. Unless it is a collectible or antique such painstaking repairs might not be justified.

Check the whole alarm mechanism. This is usually a "straight-up" assembly and any malfunction here can be detected by turning the knob that sets the alarm. It rarely poses any problem.

As in the case of the pallet pins, the correction of any other obvious faults is a matter of value judgment—weighing the cost of the clock against the time and effort involved in making the repair. Is there justification, for example, in the replacement of a broken mainspring even if one could be found? These springs are rarely housed in barrels so they would have to be wired or held with an automotive

Fig. 11-5. The arrow indicates the screw that is backed off to release the balance wheel and the hairspring.

hose clamp for removal when a replacement was found at a materials supply house. No easy matter! This is a repair that is quite possible but certainly not practical,

Remember the comic strip characters who were depicted as tossing alarm clocks out of the window? This impulse was usually generated by the alarm going off and awakening them. If the time telling portion of the movement works admirably, why not succumb to the temptation on occasion to remove the alarm mechanism (with its broken spring) and set the clock up once again as a simple timekeeper? After all, Aunt Harriet might have loved it and it still works!

An alarm clock is a strictly functional item. It does what it was designed to do most efficiently. The fact that it will run slower as the mainspring unwinds is usually of no serious concern because it is rewound and reset every single night of the year. Once it accomplishes its task of awakening someone in the morning, it is shut off imnediately and promptly forgotten. It is not expected to run accurately for more than 12 hours and, in the case of travel alarms, might lie hidden away in a bureau drawer months on end before being pressed into use for a weekend trip.

In essence, it will be found that these mass-produced items cannot be faulted for premature failure anymore than any other product. Breakage is usually induced by dropping or they are often put out of use by the loss of one of the winders! In many instances failure occurs as a result of an accumulation of dirt. We encountered one that had come off the assembly line with the dust built into it! It stopped running two weeks after it had been purchased. The owner railed against the poor quality of *all* products manufactured these days and, curious, I asked to see the clock. The movement did not even have to be removed from the case to reveal a tiny bit of fluffy dust lodged between the leaves of a pinion and the teeth of a wheel that meshed with it.

Still another malfunction worthy of mention is that caused by the clock's being overwound. The normal repair procedure is to remove the case and to manipulate the click to the position where it is free of the ratchet wheel, while holding the winding key. Then, carefully, very carefully, the pressure on the key is relaxed so that the spring unwinds slowly. It does not have to be let down all the way, and the click should be returned to its normal position frequently so that your fingers on the key can be repositioned for a strong grip. If the key slips, at this point damage could occur to the clock.

Sometimes the clock can be set running again by the simple expedient of holding it in the right hand and rotating it gently in a clockwise direction

Fig. 11-6. The end of the mainspring (arrow) is hung on a projection on back plate rather than on one of the pillars (the usual procedure).

Fig. 11-7. A method of retaining the mainspring if the holder or small hose clamp is not available.

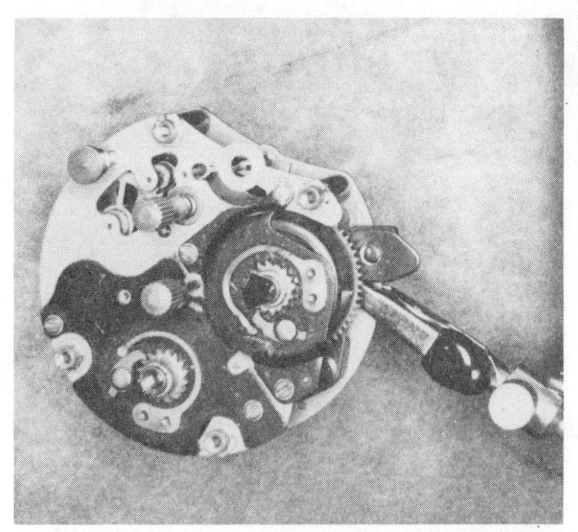

Fig. 11-8. A movement from a little desk alarm clock. Note the bent click spring (arrow).

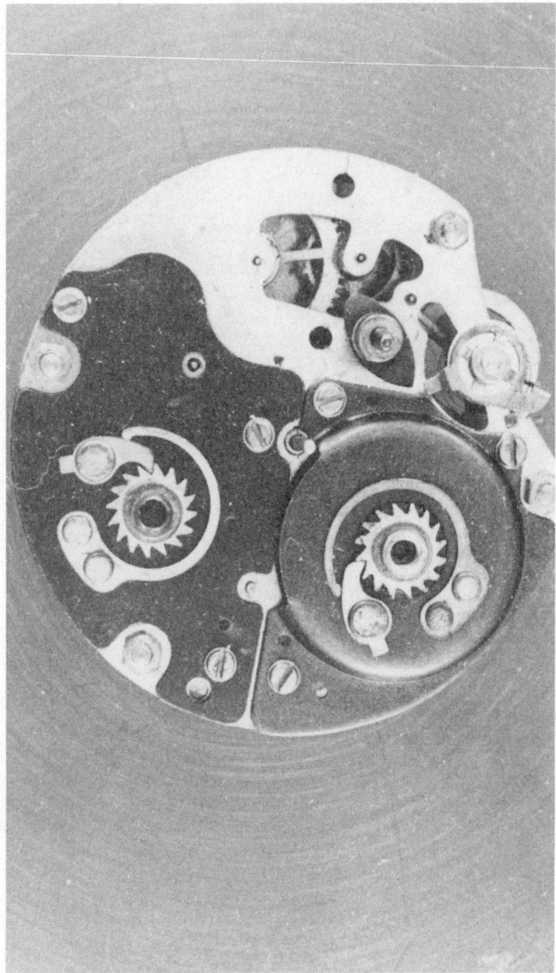

Fig. 11-9. This movement supplied good click springs and other pieces to make a complete, working model desk alarm.

and then back in the opposite direction. This will activate the balance wheel. Repeating this procedure several times over a period of five minutes works more often than not. It is certainly worth a try before you resort to disassembling the clock.

At the other end of the spectrum are the expensive little alarm clocks possibly made for use on executives' desks. With brass plates, both springs

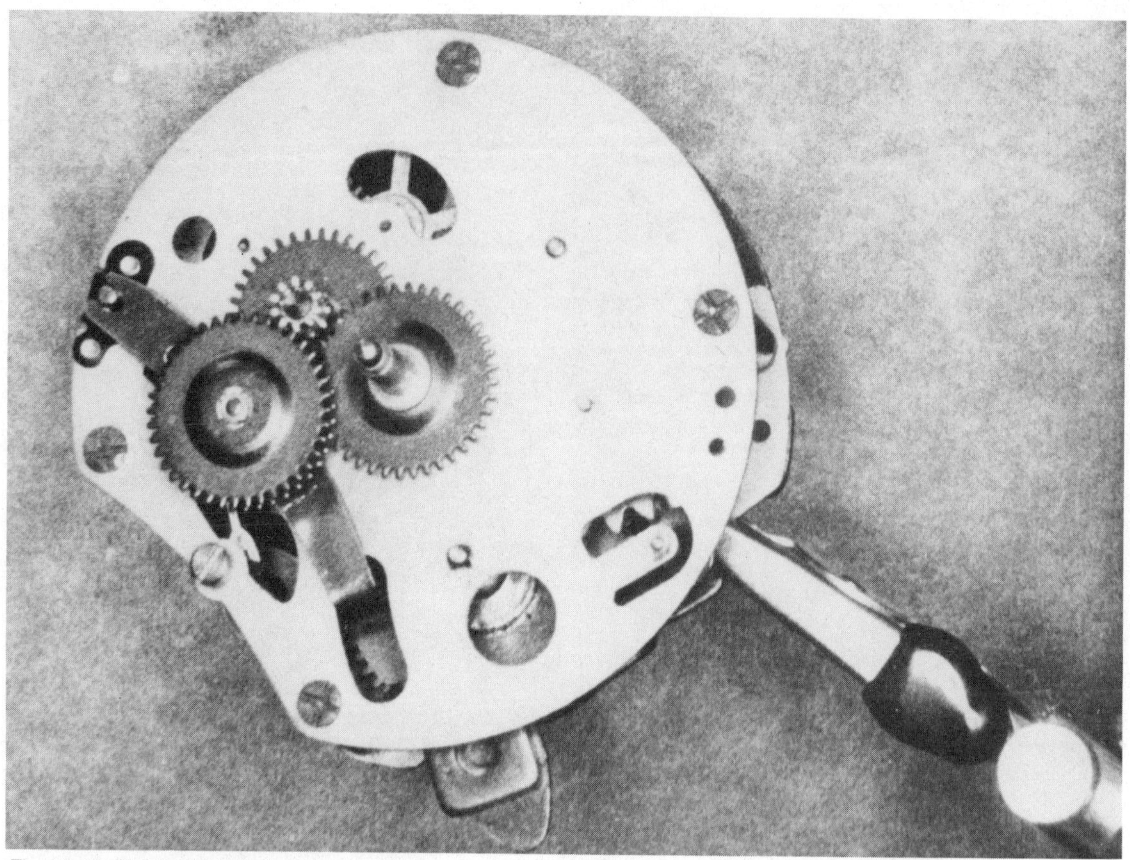

Fig. 11-10. The front view of a miniature desk alarm clock movement.

in barrels and elaborate click-and-ratchet mechanisms, they just ask to be restored or set working again. I discovered one of these, with almost solid brass plates less than 2½ inches in diameter, in a batch of junk. The retaining springs for the clicks were bent so badly as to be irreparable (especially since one end had been riveted to the plate).

It seemed like such a fine example of craftsmanship—whether mass produced or not—that I laid it aside rather than discard it. Imagine my delight when, more than a year later, I came across another almost identical one. By switching parts, I was able to produce one whole specimen complete with a working alarm and the necessary winders. No decision has been made yet about what to do with it, but it is a joy just to look at once in a while.

If repairs cannot be made, do not toss the clock out of the window. The wheels, hairspring, and some of the other components might prove useful some time in the future. Many of the stamped, brass wheels and even the pinions can be used to enhance the appearance of a scratch-built clock. A well-stocked parts bin can be a vital adjunct to successful clock building, repair, and maintenance.

Chapter 12

Special Repairs

FITTING TEETH TO A WHEEL IS DISCUSSED IN Chapter 8. The repair was slipped in there because it tied in with the work being discussed at that time. You might not be called upon to repair broken hands very often because materials and parts dealers offer a large selection of hands in all sizes for replacement purposes. Some catalogs even show the actual size of the hands and in silhouette. Frequently, all that is required to substitute a hand is to grind out the hole through which it is fitted to the arbor. In the case of some imported clocks, however, replacement hands are not available via mail order.

When the hand breaks off close to the center, the usual recommendation is to select a piece of thin steel that is just as thick as the hand and drill a hole in it that is then enlarged to the proper inside diameter. This piece is filed up to match the broken hand. This is a tedious method and for years now I have retained the center piece and used it as part of the new hand. A hand, matching the broken one as closely as possible, is selected from among those saved religiously from disassembled clocks. The rounded end that fits on the arbor should have an outside diameter larger than that of the original hand (if possible). It is cleaned thoroughly (all old paint is removed) and the original, broken off center is epoxied on beneath it. The hole is then filed out to fit the arbor. If the original hand had a collet in the center of it, this can be retained and used with the new structure.

RACK-AND-SNAIL STRIKING

Little emphasis has been placed on the repair of the rack-and-snail striking system (the one fitted to cuckoo clocks). The mechanism, usually hung on the front plate and coupled to the motion work, does not malfunction that often. Trouble here can usually be traced to worn parts or in some instances nothing more serious than a lack of oil. The system is built up of layers that can bind on the posts on which they move. Any overly tight lever could be hard to find, but forewarning about the potential problem here can help in detecting an improper action. Congealed oil can be a troublemaker here as well as in other portions of the movement.

DIALS

This is an area that should be approached gingerly in the event that restoration is necessary. There is nothing more disconcerting than to have the marks rub off during an attempt to clean one. On older valuable timepieces, particularly when not too much damage has been done over the years, it is a good idea to leave the dial alone. Some brass dials can be restored if not too deeply scratched but, in general, I've resorted to the old clockmaker's method of placing the brass disc on a lathe and sanding away all of the markings with the scratches. Using progressively finer grits of sandpaper will produce as fine a sheen as can be desired. New markings can then be designed and applied to the disc (described in Chapter 5).

Spandrels can be replaced through materials suppliers or even made up in a number of ways. Paper dials can be photographed or Photostated after being retouched and the final product will prove most satisfactory. It is well nigh impossible to repair glazed enamel on metal dials, and if a replacement cannot be obtained the solution might be to resort to photography once again.

CASES

A number of books could be written on the subject of the repair of cases. The variety of materials, the different techniques, and the shapes all contribute to the suggestion that their repair might require the aid of a specialist or that research be done on each individual case that is to be tackled. Wood seems to have predominated in clock-case fabrication over the years and, fortunately, this is a material with which most people are familiar. A good rule-of-thumb would be that the older the case, the more intricate the detailing on it, the more imperative it is that outside expert advice be sought.

Cases with raised veneer can be repaired by the layman knowledgeable enough to prick a raised bubble and press the veneer back into place with a warm iron. Veneer that has pulled away from a case can be glued back into place without too much difficulty. Most repairs here require techniques brought over from woodworking on which there is a wealth of printed material. Just perusing catalogs from suppliers like Constantine and The Woodworker's Store will offer valuable clues, and concrete help can be obtained by writing to them.

Many repairs will require nothing more than good, common sense or judgment and some few tools. This applies to metal cases, also, and to the restoration of the paint with which they may have been covered. Much information about the newer materials, the acrylics, etc., can be gleaned from Chapters 4 through 7. On pain of being castigated by the devotees of tall-case clocks, I tend to relegate the restoration of these and similar cases to the realm of woodworking. But then did not the great masters involve themselves totally with the movements and farm out the case work?

Mention was made earlier of having cleaned a marble case with soap and water. It has been suggested that a metal polish or an automobile polish (both of which are abrasive) will prove more effective despite the fact that a portion of the surface will be removed. Gilded metal should get the soap and water treatment and nothing stronger.

One portion of the case that should never, ever be neglected is the bezel. It does so much to bring the eye to the most important part of the clock—the dial. Frequently made of brass, they are generally highly polished and lacquered or varnished. Sometimes a gentle cleaning with detergent soap and water will "bring them up" and restore them to normal. If the varnish has worn through, all of it should be removed, the bezel polished well (scratches removed, etc.), and revarnished. Replacements for broken glasses in front of the dial, especially the convex ones, should be purchased from a materials supplier. Glass domes for anniversary and similar types of clocks are available from the materials suppliers. Most attractive six-sided covers can be made up from acrylic, and any endeavor in this direction is certainly well worth the effort.

Chapter 13

Maintenance

MAINTENANCE, WHEN SEPARATED FROM-repair, turns out to be such a nebulous, gray area! Clocks, unless affected by some external disturbance, tend to run on and on ad infinitum. Temperature variations do have an effect on some movements. Permanent placement in an area exposed to a large amount of sunlight can have a devastating effect on the case, especially if it is made of wood. Keeping the clock clean will add to its longevity. No matter how carefully the case is built, how it is fitted, dust will seep into the movement. When parts mesh continually, tiny little particles are removed from the surfaces through friction. Oil, introduced into the movement to reduce friction, picks up these particles and goes to work as an abrasive compound. The answer to this is periodic cleaning—at least every three years. A careful inspection at this time for any indication of wear (pivot holes, pinion leaves, etc.), is highly recommended. Cleaning a clock is not a difficult operation. Clock-cleaning solutions are readily available from the materials suppliers. A perfectly acceptable substitute for commercial products can be made up of a mild detergent soap in warm water to which ammonia is added. Many of the old authorities used to recommend gasoline and benzine (the latter is the better of the two), but these are so dangerous as fire hazards that I can only recommend not using them. It is said that laymen resort to using kerosene (poking it between the wheels on the end of a feather without disassembling the clock). Needless to say, such an operator was not prepared for what the kerosene did to the dial!

The movement or parts to be cleaned must be immersed totally in the bath to prevent marks from appearing on the portions not submerged. The bath removes the oil, but just loosens the dirt so brushing is always recommended. After the bath, a rinse, and thorough drying, all the brass parts must be polished with a metal polish such as Noxon. These metal parts can also be buff polished if they are to be seen as in a skeleton clock. Pivot holes are generally cleaned last by twirling toothpicks in them until the toothpick comes out clean. Teeth on wheels and pinion leaves can be cleaned with brass brushes in devices like the Dremel Moto-Tool. They come up gleaming after this treatment, but the pinion leaves require the use of toothpicks to clean them completely.

Enough cannot be said about using oil sparingly. It is difficult to resist putting it on all the wheels instead of just the pivot holes, the mainspring and the teeth of the escape wheel (and every other one at that!). Krylon, a subsidiary of the Borden Co., makes a product called Silicone Spray (No. 1325) that it recommends for lubricating clocks, timers, and electronic devices. It works miracles when sprayed into noisy alternating-current movements. Sprayed liberally into a movement placed on its side so that it can drain, this substance will flush out the dirt and grime in seconds and, in an emergency, can eliminate the "teardown" as a step in maintenance. Porpoise oil can still be considered by some as best for lubricating clocks, but you should not negate the importance of new products (some of them true breakthroughs) that are being presented to the public every day. Clocks treated with Silicone Spray have been running for more than two years in my studio.

Many of the other procedures that would normally fall under the heading of this chapter have been discussed repeatedly in previous sections. Any clock on which the hands are set from the front should be wound regularly to avoid having to touch the hands any more than is absolutely necessary. Any clock that functions best when in a perfectly level position should be moved as little as possible. A periodic inspection of the line cord on ac-powered movements should be made with particular attention paid to the points where it enters the wall plug, and also where it exits from the back of the case. These are areas where electric shorts are most prone to occur.

As a general rule, wooden cases can be treated almost as if they are furniture. They should be waxed to preserve them and then dusted frequently. Areas of high humidity should be avoided when setting them up on a permanent basis, They should never be positioned close to a hot-air register, a fireplace, or an air conditioner.

A point to remember with regard to grandfather clock cases is that they were built to support the weight of the movement, the pendulum, and the weights on the chains. They should be maintained in a condition of perpetual rigidness. Any flexibility in them could cause even a faint moment of sway which, upon coinciding with the swing of the pendulum, would throw the clock out of beat and eventually stopping it.

Check the method by which the movement was mounted to the case. Modern clocks generally use lugs fitted to the four corners of the back plate. Screws, through holes in the lugs, hold the movement. The vibration inherent in such a movement, with its chimes and striking, tends to cause the screws to work loose or perhaps to sag in their holes. Repairs of this defect should be made immediately upon detection.

Chapter 14

Converting Mechanical Movements to Electrical Use

At flea markets, rummage sales, and even at auctions you can, on occasion, find beautiful mantel clocks with dials bearing holes either directly under the hands or in close proximity to the 8 and the 4. This, plus a line cord emanating from the back, are dead giveaways that the clock has been converted to electricity. Such a bastardization is an anathema to a bonafide collector no matter how excellent or valuable the case. Purists tend to shy away from any conversions of spring-wound or weight-driven movements. An exception is steeple clocks where such a change might be necessary for a variety of reasons.

Retaining the original movement for such a conversion, if it can be repaired or is in working order, is frowned upon also except perhaps in the case of steeple clocks. It is felt that there would be some justification for such a conversion in the case of a valuable movement with a broken and irreplaceable mainspring, particularly if it had been designed to run for 8 days or, at best, 30 days. People have been known to make such a request just for sentimental reasons.

There are several ways that this can be done—convert a movement, not switch it—without the need for mechanical engineering knowledge or more ability than that required to build any of the clocks described in this book. For the adventuresome few with mechanical inclinations and aptitude, the simplest method of adding a motor to an old clock is to take out the mainspring and any unnecessary wheels and couple the motor shaft to the center arbor.

The key to all of this would be to select a motor that puts out the correct amount of revolutions right at the shaft to be used in coupling. The size of the wheels in the clock and the number of teeth thereon would be a prime determining factor. The method of

Fig. 14-1. The prototype spring-wound movement converted to alternating current power.

coupling would entail the use of a short length of spring with a large enough diameter to fit tightly over both the arbor and the shaft from the motor. This would produce a flexible coupling. The disadvantage of this setup lies in that the motor, coupled to the back of the movement, would require an elongated case with a most unusual shape! This unusual aspect might be diminished somewhat if a helical gear was fitted to the driving shaft and mated to a wheel of proper dimensions on the driven shaft. See Fig. 14-1.

The method with which I am experimenting currently is to retain everything in the old clock, but the mainspring, its arbor, and wheel. To this is fitted a conventional synchronous motor similar to those used in Sessions, Spartus or Westclox movements. Because the pinion on the arbor of the motor usually has but 10 leaves, one or two small wheels will have to be interposed to couple it to one of the wheels in the clock.

Obviously, the normal direction of rotation of this pinion also is important. A small plate of acrylic or thin steel is made up with a hole in it to accept the pinion from the motor (which is bolted to the back of the plate). One small wheel, approximately ⅝ of an inch in diameter, in this case, is positioned immediately beneath the pinion so that its teeth mesh properly. The second wheel, of about the same diameter, is placed to the left of the first wheel so that its teeth mesh with the pinion on the first one. They are held in position with a bridge.

The whole assembly is arranged so that the pinion fitted to the second wheel meshes with the teeth on a wheel in the clock. Because the escapement with balance wheel is retained, fast or slow regulation can still be effected. In making or selecting a case for this conversion, extra room would have to be provided above the movement to allow for the positioning of the motor above the whole works. A steeple-type casing would not look out of proportion.

A small alarm clock movement was used in conducting these experiments, but these little motors are strong enough to power 8-day movements easily. The concept is sound, however, and my work with it goes on apace.

The preceding suggestions are for converting spring-wound movements to electric power. The initial idea has been used on occasion to make an electric clock. This immediately gets into a slightly different field, one to which engineers rather than laymen are generally attracted. The setups are far more complicated than anything I want to include in this book.

The English seem to enjoy this type of endeavor, invariably making a tall wooden case and a hand-made pendulum with cylindrical bob an integral part of the clock. There are a number of obscure little British books on this subject, some of which have been reprinted in this country. Caldwell Industries (address in the Appendix of this book) lists some of them as does the Adams Brown Company (hولological literature). A diligent search in large libraries should turn up still others.

Chapter 15

Engraving and Etching

MUCH MENTION HAS BEEN MADE THROUGHout these pages of engraving and etching. These arts (or crafts) owe their origin to the age-old desire to cover a multitude of flaws inherent in any given material. A prime example of this is the difficulty of working with a sheet of brass, copper, silver, or steel without getting scratches on it. Of even greater importance is the fact that these metals will not retain permanently any overlay (paint, enamel, lacquer, etc.), unless some portion of the addition is laid on in minute troughs or canals just below the surface. It was discovered centuries ago that, in many instances, this "grooving" was, in itself, esthetically pleasing to the eye. Such techniques, stippling on glass, for example, will enhance the appearance of what normally would be considered an inferior material. It was in pursuit of just such effects that ladies, during the Victorian era, spent hours on end tapping little dots on the surface of glass.

A high percentage of the engraving employed on the projects in this book serves a truly functional purpose over and above embellishment. Numerals, batons, lines, spandrels, and cartouches can all be done this way. Clock plates of metal or plastic take on a new look when, without the treatment, they might have been considered unserviceable.

There is no intent here to diminish engraving as a fine craft in its own right. There are people who have spent years perfecting skills with scribers, gravers, die sinkers, chisels, and knives. Often this skill (plus God-given talent) had been dedicated to the beautification and embellishment of one material—glass, metal or, more recently, plastic. Many true artists in this field could never have honed their skills to a fine degree of perfection had they spread their interests to the building of clocks, firearms, the tooling of leather, the making of jewelry, watches, and furniture. Conversely, the realization of full satisfaction from any of these crafts would suffer if a large amount of time was placed on the engraving.

What is suggested here is that the techniques for the necessary engraving on clock dials, plates, and portions of cases be held to a small portion of the overall project because of the time involved. It is for this reason that I have embraced the use of various electric engravers of rotary and impact design. Their efficiency, the speed with which they produce a design, and the ease with which they can reproduce accurately make them eminently worthy of inclusion in this book.

Acrylics are exciting because they can pipe light. Glass, with some light-piping capabilities, is far more difficult to engrave. Etching proves most effective with glass. The formulation of some new etching pastes with reduced toxicity have spurred a resurgence in the popularity of this art, and its use as part of clock building can be regarded as a bonus to builders.

Commercially available, electric-powered engraving tools break down into two basic categories: the impact type and the rotary type. The former usually consists of a piston that is thrust forward and drawn back with amazing rapidity. The piston bangs upon the end of a rod to the other end of which

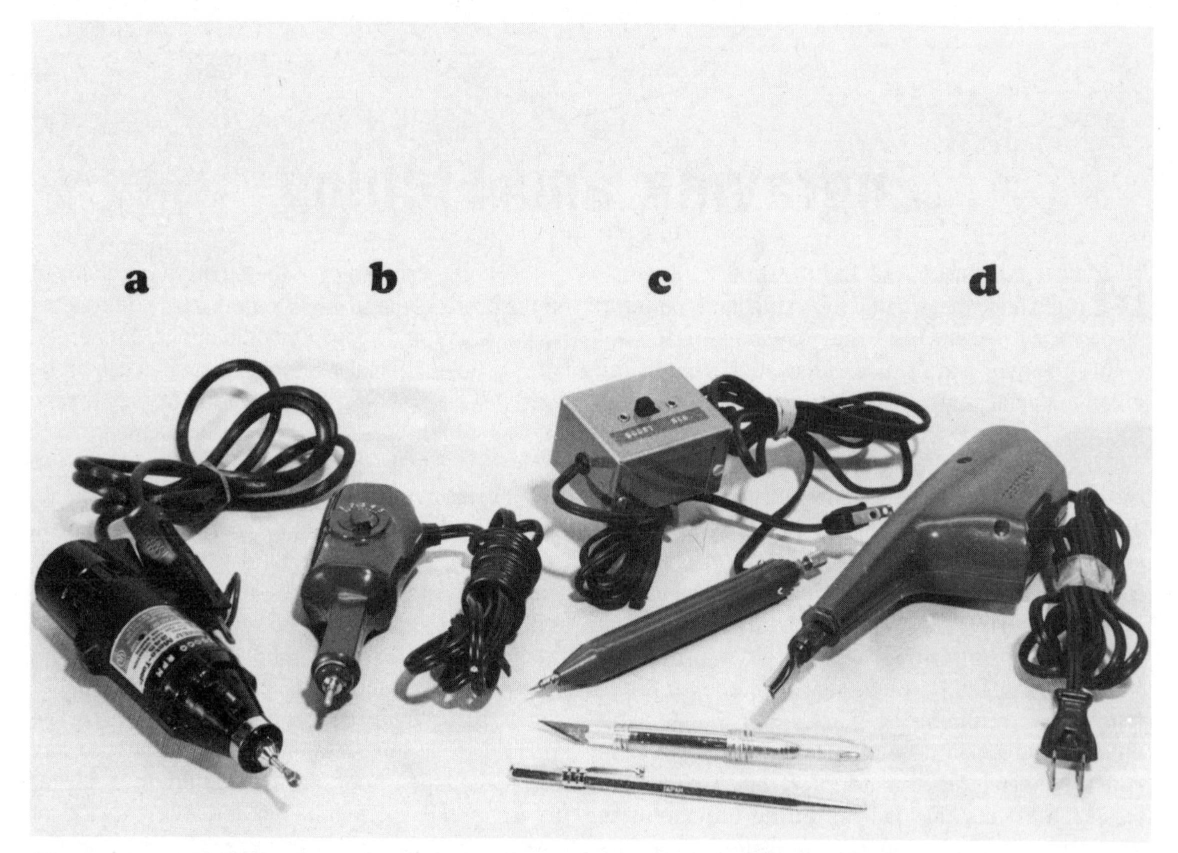

Fig. 15-1. A variety of engraving tools: (A) Dremel Moto-Tool; (B) Dremel Electric Engraver; (C) Electro-stylus; (D) Craftsman Electric Vibrator. Front center: X-Acto knife and carbide-tipped, pencil-type stylus.

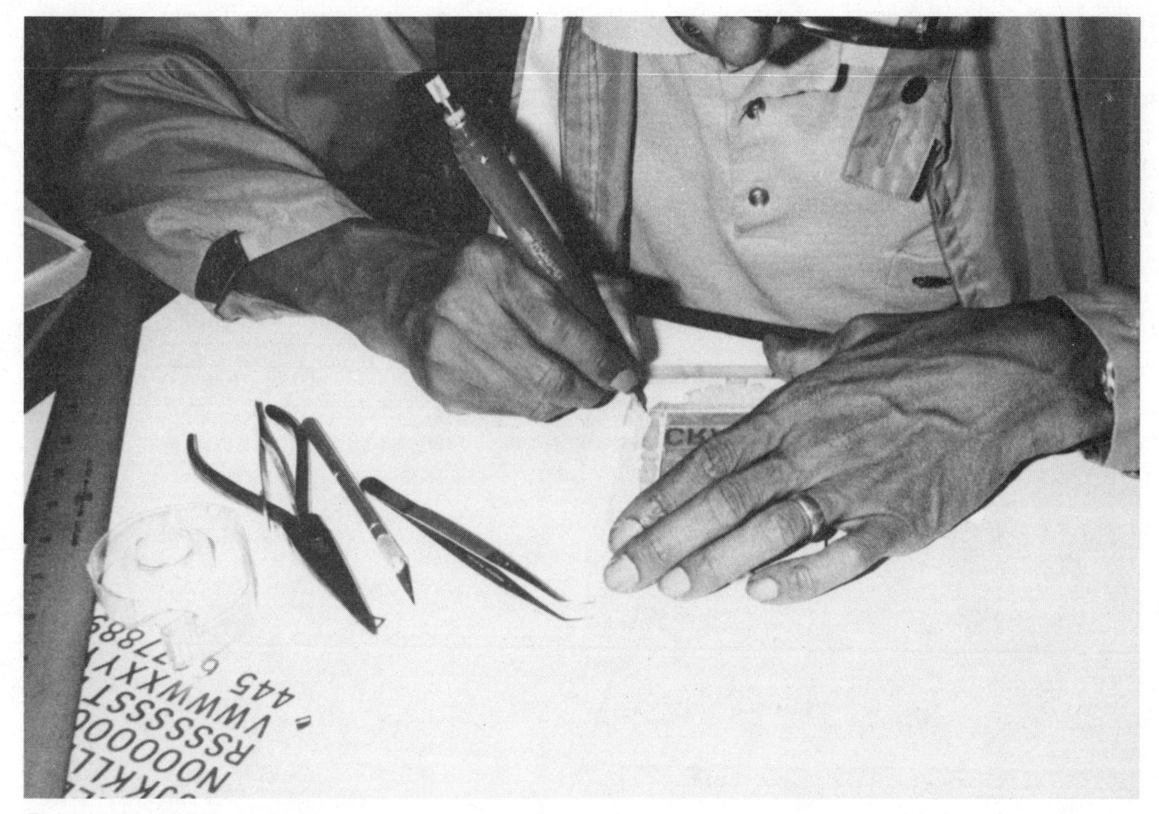

Fig. 15-2. Engraving with the Electro-stylus.

is fitted a pointed tool or, in some instances, a modified chisel point. These tools work very well with little practice, but they are tiring when used for long periods of time due to the loud noise they make.

One such device sold commercially under the name of Electro-stylus (Fig. 15-1) was used to do a good portion of the engraving shown within these pages. See Figs. 15-2 through 15-6. In addition to the noise level (not quite as high as some of the others), its main drawback lies in the fact that its casing tends to get hot. After an inordinately short time in use, it must be shut off and allowed to cool down.

The force of the stroke can be varied by adjusting a knurled knob on one end. While the adjustment will make the tool produce a fine or even dotted line, it does lengthen the running time between cool-down periods. It is a good tool and would be used a lot more were it not for this failing that no amount of experimenting by me has been able to eliminate.

Sears, Roebuck and Company makes an impact-type of tool that sells well at present because it is excellent for marking tools and items of value for identification in the event of theft. It also has a high noise level that makes it unsuitable for use over extended periods of time. Called the Craftsman Electric Vibrator, it comes equipped with an engraving point, saw and knife blades, chisels, gouges, and an abrasive point. The engraving point, of most concern here, marks permanently on wood, steel, plastic, glass, aluminum, copper, brass, stone, iron, and other surfaces. The abrasive point is said to be suitable for frosting glass.

Dremel Mfg., specialists in this type of tool, offers an Electric Engraver quite similar in concept to the Craftsman except that it is a little more compact and lighter in weight. The depth of engraving is controlled by a five-position knob on the side. I recommend using the lowest setting because it will produce a deep enough engraving. The engraver point furnished with the tool is solid carbide steel, and it is adequate for the work to be done in connection with clocks. For continued operation, nevertheless, a diamond point is recommended by the manufacturer.

Another well-known impact engraver is the Burgess Vibro-Tool. It offers a choice of carbide-tipped, solid-carbide or diamond points. For most metals the carbide-tipped or solid-carbide points are recommended. The diamond point is recommended for continuous engraving and for work on very hard metals. For glass, plastics, and ceramics, the technique is similar to metal engraving except that the point is held perpendicular to the surface to avoid chipping. In this instance, adequate cushioning is recommended for the glass and ceramic items. The diamond point is suggested for continu-

Fig. 15-3. An example of fine engraving on an antique rifle (courtesy of Angus Laidlaw).

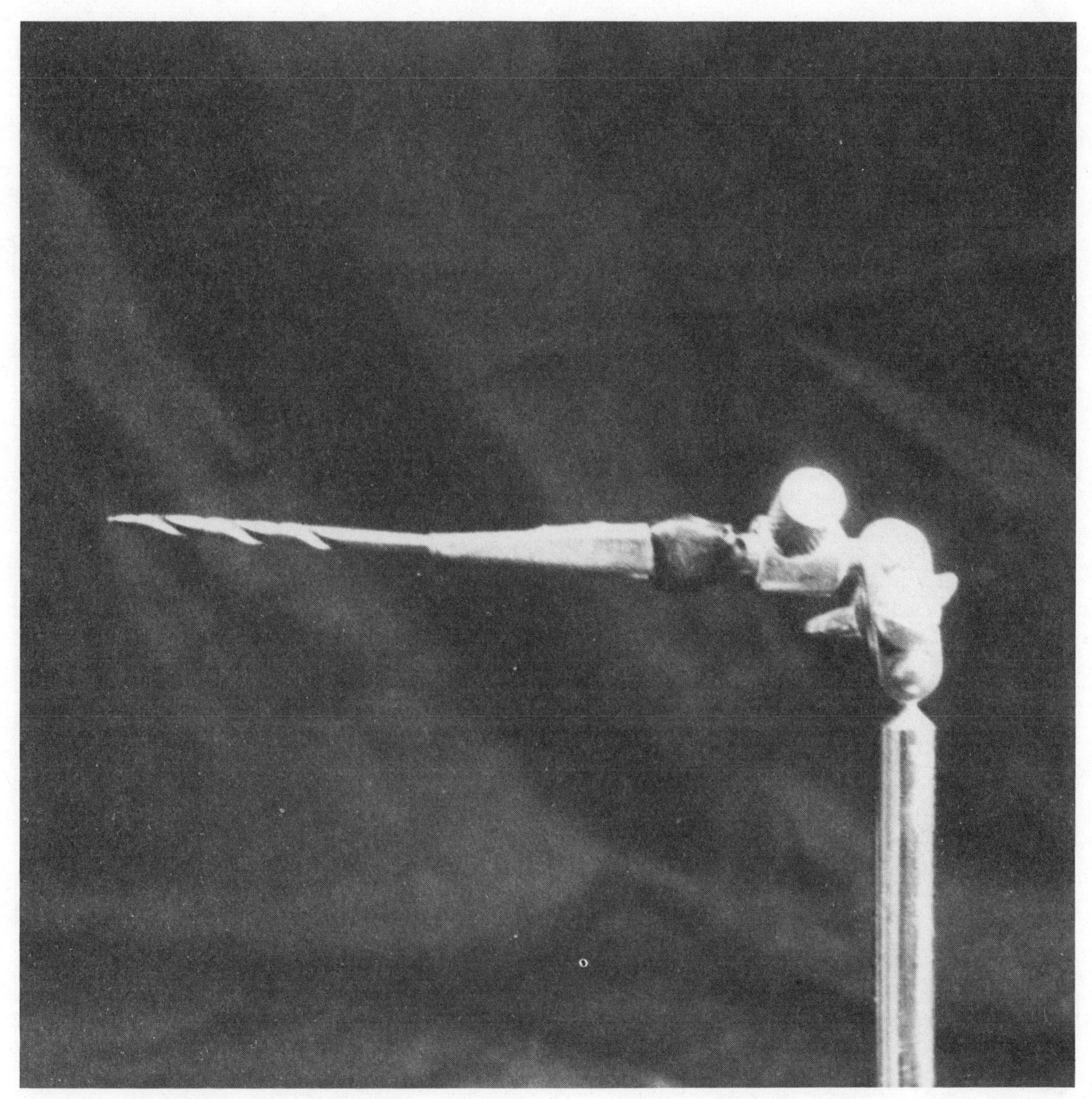

Fig. 15-4. A drill bit reground for engraving.

ous, heavy-duty jobs. Here, too—as in the case of the Dremel Electric Engraver—Burgess' abrasive ball point is suggested for frosting and shading glass.

The Vibro-Tool has a noise level of intensity comparable to the other two tools mentioned above. It also tends to heat up when used on the LO speed setting, coupled with the maximum stroke for periods in excess of 15 minutes. The manufacturer suggests that it be shut off to allow the heat, a result of the added power at this setting, to dissipate. The company states that the heat is not due to any malfunction. It is used best when held as a pencil—especially for engraving, tracing, and frosting. It does not get as much use in my studio as do the others.

The rotary engravers, of which the Dremel Moto-Tool is a good example, are far and away the most versatile of these tools. Variations on a theme are Dremel's Moto-Flex Tool and the Moto-Shop

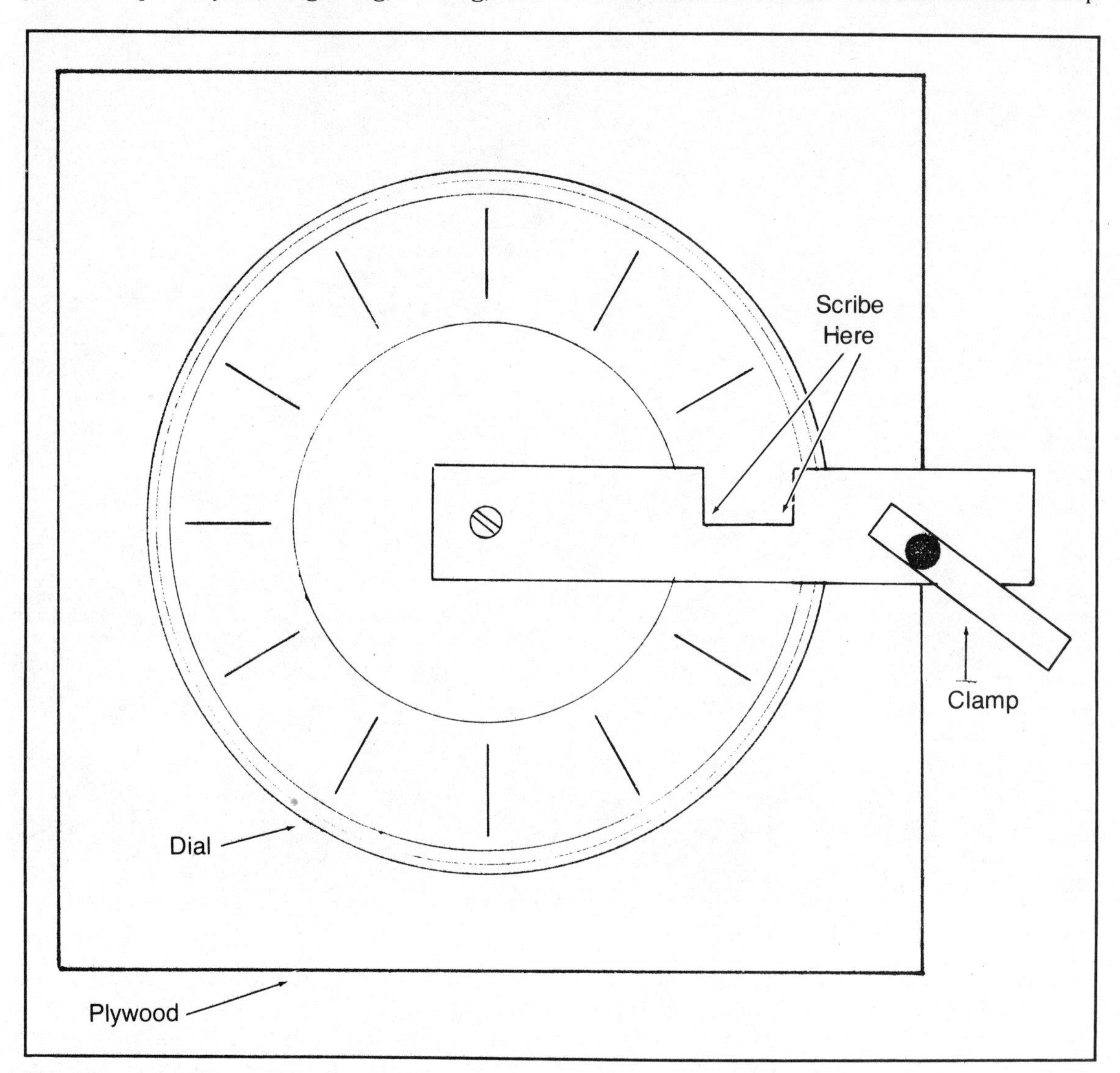

Fig. 15-5. A device for engraving radial lines.

Fig. 15-6. Foam-covered lazy Susan for engraving and other tasks where the workpiece has to be rotated on a horizontal plane.

equipped with a flex-shaft. All of them use, interchangeably, emery-wheel and silicon-grinding points especially suited for engraving.

Extremely small cutters also can be used, although the flutes tend to "load up" (fill up with ground-off material) after protracted use. The Dremel Moto-Tool, Model 370, has variable speeds from 5000 to 25,000 rpm. This tool can be used with short drill bits, to ⅛ inch or 3/32 inch in diameter, ground to a needlelike tapered point.

These tools are the ones used most frequently for the projects described in this book. The Moto-Flex Tool produces far less heat than any of the others. It can be fitted with a solid-state speed control for added versatility. The thin, pencil-like handpiece is comfortable to use and it allows for precise engraving of fine lines. Rather than use the stand, I opted to hang it from a chain in the ceiling; this keeps both the line cord and the shaft out of the way.

One of the simplest ways to get used to these tools and to see just what kinds of lines they produce is to select a scrap piece of black acrylic and spray one surface with white lacquer. The lines will stand out most effectively. Then switch to a piece of clear acrylic. The lines will still be visible, but much more so if the workpiece is set on a black surface. I have made up a lazy Susan with two discs of plywood about 9 inches in diameter, one of which revolves upon the other. Between them is fastened a ball-bearing device obtained from Edmund Scientific in New Jersey. The top disc covered with black foam, revolves smoothly even with the pressure of one's hand holding the engraver upon it. The work light used will have to be moved around until it provides just the right amount of scaling light that is correct for the individual engraver.

No freehand work is suggested here. The pattern to be transferred onto the acrylic, glass, or metal is traced onto the protective paper covering the acrylic or, with carbon paper, onto glass or metal. I have used pieces of frisket—obtainable from art supply stores—most frequently. This is a translucent, adhesive-backed film. It is laid upon the pattern, a tracing is made directly on it, and then its protective back is removed and it is pressed onto the surface to be engraved.

An impact tool is best for outlining the design and filling in the guidelines. A rotary tool can be used afterward for cross-hatching and fill-in work. Spandrels are easiest to do with an impact-type tool, but a rotary-type tool fitted with a small engraving cutter similar to Dremel's No. 108 or 109 will cut through the frisket and leave an impression without tearing it away from the workpiece.

Engraving on clear acrylic is done on the back of the piece for the best effect. Only plate glass can be done this way; tumblers, glasses, bowls, etc., are generally engraved on the outside. A striking effect can be achieved by using ¼ -inch or slightly thicker acrylic sheet and making the engraving much deeper than usual. A pleasing three-dimensional effect is produced, and especially in the case of numerals and batons on clock dials. One prohibition would be against the use of sheet thicker than ⅜ of an inch material that could present difficulties in installing threaded, center-mount fit-ups.

Engine turning can be considered a form of engraving if it is used to enhance the appearance of a piece rather than just to cover scratches. It is done most easily with a drill press. That limits it to the owners of well-equipped workshops. There is no limitation to the size of the workpiece that can be engine turned other than the capacity of the drill press used. The size of the circular brush used would depend on the area to be covered—smaller brushes for smaller areas (in square inches). It will work on acrylic though not quite as well as on metal. The pressure of the brush against the acrylic must be extremely light and a steel brush seems to work better than one with brass fibers.

The engraving tools can be used for scribing lines, but there is a manual scriber known as a cutting tool for Plexiglas acrylic sheet available from plastic supply houses that is much easier to use. In an emergency, an X-Acto knife with a #11 blade will work if the back of the blade is used rather than the cutting edge. Several strokes can be required to achieve the required depth or to accent lines radiating from the center of a dial to the numerals 3, 6, 9, and 12.

Using a lathe to scribe concentric lines on a disc has been mentioned previously. This procedure works equally well on metal or acrylic. The coarseness of the sandpaper used has a direct effect upon the depth and visibility of the lines. Once the paper is held against the disc, it should not be moved except to lift it off. Movement toward or away from the center of the disc will blur the lines,

The description of a simple device for engraving radial lines on a dial follows. Cut a backing board of plywood about 12 inches square. Drill a 3/16-inch hole in the very center of it. Cut several pieces of aluminum or tin about 7 inches long and 1-inch wide. Drill a hole in each piece centered about ½ inch in from the end. Cut a notch in the piece just where the scribed line (for a baton usually) should be made on the dial. Now just lay the dial in the center of the backing board, bolt the metal piece in the center of the assembly on top of the dial, and clamp the other end. The lines can now be scribed with the cutting tool mentioned above. See Fig. 15-5.

Glass is the prime material under consideration for etching work because plastics do not lend themselves to this process. Relatively large areas and elaborate designs can be worked onto a piece of glass in a shorter period of time than if done by engraving. It is suggested for the center of dials and for glass mirror front or side plates. There are a number of companies that manufacture the etching cream that is a prerequisite to glass etching in today's world; a product named Etchall is among the better known creams,

It is not within the scope of this book to delve deeply into the uses of these etching creams, but each manufacturer furnishes full and complete instructions, plus patterns and sometimes stencils. In brief, a stencil is carefully placed on the surface of a glass object and the etching cream is applied liberally. After a few minutes, the cream is washed off with flowing water and the stencil is removed. The pattern is permanently etched onto the surface of the glass. The effect is a very pleasant one and will enhance dials as well as the bottom portion of doors on mantel clocks. Many of the creams are caustic and toxic and the instructions should be read very carefully to determine whether or not the use of gloves is required.

Armour Products, Midland Park, New Jersey 07432, offers a blister-packed glass etching kit with stencils, a brush, and a vial of its cream—Armour Etch. It should do a number of etchings quite inexpensively! A good book on the subject including 50 patterns is available from the Chrome Yellow Private Press, P. O. Box 14082, Gainesville, Fla. 32604. The title is *Etching and Sandblasting Glass* by Nord.

My inclination is toward engraving because it can be applied to so many different materials, including the mirrored acrylic available from most plastic supply houses.

Appendices

Appendix A

Kit Manufacturers and Distributors

Albert Constantine & Son Incorporated
2050 Eastchester Road
Bronx, NY 10461
grandfather, grandmother, wooden wheel

Arrow Handicraft Corporation
900 West 45th Street
Chicago, IL 60609
miniature American wall clock

Caldwell Industries
Box 170
Luling, TX 78648
synchronomes, Matlock-Collins

Craft Patterns Studio
Route 83 & North Avenue
Elmhurst, IL 60126
grandfather, grandmother, school

Craftplans
8011 Lewis Road
Minneapolis, MN 55427
grandfather, grandmother, wooden wheel

Craft Products Company
North Avenue & Route 83
Elmhurst, IL 60126
grandfather, Lancaster (skeleton), school, traditional American, Vienna regulator

Craftsman Wood Service Company
2727 South Mary Street
Chicago, IL 60608
grandfather

H. De Covnick & Son
P. O. Box 68
Alamo, CA 94507
grandfather, grandmother, traditional American, school

Emperor Clock Company
Emperor Industrial Park
Fairhope, AL 36532
grandfather, grandmother, school, Bavarian, traditional American

General Time Service
2646 North Greenfield Road
Mesa, AZ 85201
various decorative electronic types

Heath Company
Benton Harbor, MI 49022
electronic

Held Products
9 Lakeview Drive
Farmington, CT 06032
wooden wheel

Hentschel Clock Company
P. O. Box A
Osterville, MA 02655
grandfather, grandmother, traditional American

Herter's Incorporated
Waseca, MN 56093
wooden wheel

JHS Enterprises
Box 81
Point Clear, AL 36564
cuckoo

Kits'n Crafts
205 Appleton Street
Holyoke, MA 01040
grandfather, decoupage, school, Swiss chalet, English bracket, traditional American

Klockit
P.O. Box 609
Lake Reneva, WI 53147
West German skeleton, various types

Lindberg Products, Incorporated
8050 North Monticello Avenue
Skokie, IL 60076
plastic mantel clock

Mason & Sullivan Company
586 Higgins Crowell Road
West Yarmouth, MA 02673
grandfather, grandmother, school, traditional American, English carriage, skeleton

Merritt's Antiques, Incorporated
RD #2
Douglassville, PA 19518

Minnesota Woodworkers Supply Company
925 Winnetka Avenue North
Minneapolis, MN 55427
grandfather, wooden wheel

Tom Paul Enterprises
P. O. Box 1966
Daytona Beach, FL 32015
various types

Science & Mechanics Publishing Company
Craft Print Division
299 Park Avenue South
New York, NY 10003
grandfather

Selva-Borel German Clock Catalog
347-13th Street
P. O. Box 796
Oakland, CA 94604
various types, *Black Forest clock, old German, old Swiss*

Stanley Tools
P.O. Box 1800
New Britain, CT 06050
grandfather, grandmother

Tandy Leather Corporation
(see local telephone listing)
various leather-faced types

Technical Writers Group
Box 5994 State College Station
Raleigh, NC 27607
digital

Thomas Woodcraft
1412 Drumcliffe Road
Winston-Salem, NC 27103
wooden wheels

Charles E. Vautrain Association, Incorporated
117 Park Avenue
West Springfield, MA 01089

Walnut Hollow Farm
Route #2
Dodgeville, WI 53533
rustic woodcrafted line

Westwood Clocks 'N Kits
3210 Airport Way
P.O. Box 2930
Long Beach,
CA 90801-2930

grandfather, grandmother, school, English bracket, European wall, anniversary, traditional American, Vienna regulator, German striking skeleton

Yield House
North Conway, NH 03860

cuckoo, grandfather wooden wheel

Appendix B
Suppliers

Adams Brown Company *books*
P.O. Box 357
Cranbury, NJ 08512
609/799-2125

Allcraft Tool & Supply Company, Inc.
100 Frank Road
Hicksville, NY 11801
516/433-1660

Almac Plastics Inc.
47-42 37th St.
Long Island City, NY 11106

Arlington Book Company
P.O. Box 327
Arlington, VA 22210
202/296-6750

Art Plastics
359 Canal Street
New York, NY 10013

Cadillac Plastic and Chemical Co.
35-21 Vernon Boulevard
Long Island City, NY 11106

Caldwell Industries *books, kits*
Box 170
Luling, TX 78648
512/875-2654

Canal Plastic Center
345 Canal Street
New York, NY 10013

Canal Street Plastic Supply Co.
2107 Broadway (73rd St.)
New York, NY 10023

Cas-Ker Company — *books, movements parts, tools*
P.O. Box 2347
Cincinnati, OH 45201
513/241-7075

Clock Component Center — *movements, parts*
Box 456
Belleville, MI 48111
313/699-9000

ClocKraft — *movements, parts*
2976 Bonnie Lane
Pleasant Hill, CA 94523

Clock Repair Center — *movements, parts, tools*
P.O. Box 65
Carle Place, NY 11514

Clock Repair Center — *movements, parts, tools*
220-17 Jamaica Avenue
Queens Village, NY 11428

Commercial Plastics & Supply Corp.
630 Broadway (north of Houston St.)
New York, NY 10011

Cosmopolitan Watch Material Importing Corp. — *parts, tools*
Manfred Tauring
Suite 700-709
87 Nassau Street
New York, NY 10273
212/227-9276

Craft Products Company — *books, kits, movements, tools*
Order Department
North Avenue & Route 83
Elmhurst, IL 60126
312/832-6287

C.R. Clock Shop — *movements, parts*
11906 Q. Drive North
Battle Creek, MI 49017
616/965-8585

DRS, Incorporated — *Supplier to the trade*
15 West 47th Street
New York, NY 10036
212/757-7370-1

Emerson Books, Incorporated
Reynolds Lane
Buchanan, NY 10511
914/739-3506

Emperor Clock Company — *kits, movements*
Department 04
Emperor Industrial Park
Fairhope, AL 36532
205/928-2316

H. De Covnick & Son — *kits, movements, parts*
P.O. Box 68
Alamo, CA 94507
415/837-1244

Hentschel Clock Company — *kits, movements, parts*
P.O. Box A
Osterville, MA 02655
617/428-3505

Horolovar Company — *catalog cost: $1 anniversary clocks, parts, movements, books*
Box 400
Bronxville, NY 10708

Industrial Plastic Supply Co.
309 Canal Street
New York, NY 10013

Innovation Specialties *kits, movements, parts, plans*
Dept. PCM
3410 South La Cienega Blvd.
Los Angeles, CA 90016

Kits'n Crafts *kits, movements*
205 Appleton Street
Holyoke, MA 01040

Klockit *books, movements, parts, kits*
P.O. Box 629
Lake Geneva, WI 53147
414/248-1150

The Letter King *movements, parts*
716 San Antonio Road
Palo Alto, CA 94303
415/494-2162

Mail Order Plastics
302 Canal Street
New York, NY 10013
212/226-7308

Mail Order Plastics
56 Lispanard St.
New York, NY 10013

Mason & Sullivan Company *kits, movements, parts, tools*
586 Higgens Crowell Road
West Yarmouth, MA 02673
617/778-0475

Merritt's Antiques, Incorporated
RD #2 *books, kits, movements, parts tools*
Douglassville, PA 19518
215/689-9541

Bob Morgan Woodworking Supplies
1123 Bardstown Road *woodworking, supplies, veneers*
Louisville, KY 40204
502/456-2545

National Artcraft Company *movements*
23456 Mercantile Road
Beachwood, OH 44122
216/292-4944

Tom Paul Enterprises *movements, parts, kits*
P.O. Box 1966
Daytona Beach, FL 32015
904/255-8758

Plastics Center of New York
300 Park Avenue South (22nd St.)
New York, NY 10010

Plymouth Watch Material Company, Inc.
140 Nassau Street *clock parts, clock tools*
New York, NY 10038
212/267-4441

PM *movements, parts, tools*
Modern Technical Tools & Supply Company
211 Nevada Street
Hicksville, NY 11801
516/931-7875

Primex *movements, parts*
720 Geneva Street
Lake Geneva, WI 53147
414/248-2000

Roger Quality Castings, Incorporated
Roger Metal Craft Division *parts*
Warwick Street
Boyertown, PA 19512
215/369-1304

Selva-Borel German Clock Catalog
347-13th Street *kits, movements, parts, tools*
P.O. Box 796
Oakland, CA 94604
415/832-0355

Turncraft Clock Co. *quartz battery movements, parts*
Dept. PCM
611 Winnetka Ave. North
Golden Valley, MN 55427

Charles E. Vautrain Association, Incorporated *kits*
117 Park Avenue
West Springfield, MA 01089
413/785-1551

Westwood Clocks 'N Kits (California Time Service) *kits, movements, parts*
3210 Airport Way
P.O. Box 2930
Long Beach, CA 90801-2930
213/595-5411

Ralph Woods *tools*
Box 1079
Lebanon, MO 65536
417/532-4487

Glossary

abrasive—A substance used for grinding and polishing. These are generally oilstone, emery, carborundum, lavigated aluminum oxide.

addendum—That portion of the tooth of a wheel or pinion that extends beyond the pitch circle.

alarm—The mechanism attached to a timekeeper whereby, at any set time, a hammer strikes rapidly on a bell for several seconds.

alloy—A mixture of two or more metals.

amplitude—The amount of arc or swing of the balance. This can also be measured by certain types of oscilloscopes.

analogue—Terms used to denote any timepiece with dial and hands, as opposed to digital display.

anchor escapement—The escapement used in most domestic clocks.

anchor (verge)—A device that regulates the speed of rotation of the escape wheel.

annealing—The act of heating and slowly cooling a metal or substance to render it softer or to relieve internal stresses.

Arabic figures—Figures on a dial, such as 1, 2, 3, as opposed to Roman numerals such as I, II, V, IX.

arbor—An axle, steel shaft, or rod on which wheels and pinions are affixed; commonly referred to as the barrel arbor, pallet arbor, winding arbor.

Arkansas stone—A white, marblelike stone used in various shapes and sizes as a grinding stone to sharpen gravers and tools. Used in powdered form as an abrasive with grinding laps.

armature—The iron component that moves as a result of magnetic attraction, particularly the rotor of an electric motor and the moving part of an electric bell, buzzer or similar arrangement.

balance spring—Often called the hairspring. A long, fine, spiral spring that determines the time of swing of the balance. One end is fastened to a collet that is attached to the balance staff and the other to a stud which is attached to the plate.

balance staff—The axis or arbor that carries the balance wheel.

banking pin—Two pins that limit the motion of the pallet fork. They also control the amount of slide of the pallet pins.

bar movement—Often called a bridge-model movement. One plate is omitted and its place is taken by a number of bars extending over the movement and carrying the upper pivots.

barrel—A hollow cylinder, usually of brass, turning on an arbor. Mainsprings are often contained in barrels and the line of a long-case clock is wound around one. The chiming barrel is a cylinder with pins or other projections for operating chime hammers.

barrel arbor—The axle of the barrel around which the mainspring is coiled.

barrel hook—A hook or slot in the inside of the barrel wall upon which the last coil of mainspring is attached.

beat—The tick of the clock or one swing of the balance wheel. *In Beat*: when one beat or *vibration* of the balance receives its impulse at the same distance from the line of centers as the other.

bezel—The ring holding the glass of a clock. Also the ring without a groove for a glass.

bluing—To change the color of polished steel by heating it to approximately 540° F.

bob—The ball or weight at the lower end of a pendulum.

brace—The hook or connection attached to the outer end of the mainspring.

bridge—A bar, usually supported on both ends, for holding the pivots on arbors longer than the others in a movement. This device is called a cock when it is supported on one end only.

broach—A tapered steel tool with flat cutting edges used to enlarge holes already drilled.

brocot suspension—The method of suspending a pendulum so that it can be raised or lowered by means of a key from the front of the dial.

brush—A springy contact resting on the commutator in electric motors and conveying current to a wire-wound armature. Clock synchronous motors do not have wire-wound armatures or brushes.

burnisher—A hard polished piece of steel used to polish softer metals by rubbing it along the surface to be finished.

bushing—A brass or other hard material tube inserted into a pivot hole of a clock plate to correct for wear and to allow for added wear. Block bushings amount to substitute holes that are fitted carefully to receive the pivots and then riveted into place in the movement plates.

caliper—An instrument with two adjustable curved legs used to measure diameters.

cam—A small flat piece used to transfer circular motion into back-and-forth motion to a lever or other contacted piece.

cannon pinion—A thin, steel tube with pinion leaves at its lower end and usually carrying the minute hand at its upper end.

carbide—The type of cemented carbide tools used as gravers to cut very hard metals.

carborundum—A carbon silicon substance used as an abrasive.

castle wheel—The clutch wheel.

center-seconds hand—A long second hand moving from the center of the dial and not from a separate center over a different part of the dial.

chapter—The ring on the dial plate on which are painted or engraved the hour numerals and minute gradations.

cleaning solution—A dirt, tarnish- and grease-dissolving liquid composed mostly of ammonia, oleic acid, and water.

clepsydra—A water clock.

click—The pawl used to prevent the ratchet wheel

from turning backward after the mainspring has been wound.

clickwheel—The rachet wheel on a mainspring winding square that, with click and clickspring, prevents the mainspring from unwinding save by turning the gear train.

clockwise—The direction of circular motion going in the same direction as the hands of a clock; circling from horizontal left, upward, around, and down toward lower right.

clutch—The friction drive arrangements by which the hands and their associated wheels can be turned independently of the main gear train for the purpose of setting clock hands to time.

cock—A bracket; particularly the bracket by which balance wheel or pendulum clock pallets are mounted.

collet—A brass collar or washer that holds a wheel on its arbor.

commutator—Segments of a cylinder on the armature of an electric motor by means of which, in conjunction with the brushes, current to the armature can be switched on and off according to its position in relation to the field magnet.

concave—A surface curved inward as the inside of a bowl.

convex—A domed surface as the top of a curved clock glass.

countersink—A chamfered or concave cut.

crocus—A red-brown powder used for quick polishing, slightly coarser than jewelers' rouge.

crutch—The forked lever that connects the pallets (on whose arbor it is attached) to the pendulum rod in a pendulum clock.

crutch wire—A wire that carries the impulse from the escapement to the pendulum.

dead-beat escapement—An escapement in which the escape wheel remains motionless after it becomes locked upon the pallets, as opposed to the recoil escapement.

dial—The face of a watch or clock upon which the numerals are placed to indicate the time.

dial plate—A plate, usually wood, brass, iron or plastic on which the dial is engraved or painted.

dial train—The train of wheels under the dial that motivates the hands. The cannon pinion, hour wheel, minute wheel, and pinion.

digital display—Any timepiece where time is read with digits.

diode—An electronic valve used in battery movements to suppress the spark when contact is broken.

discharging pallet—The exit pallet pin. The pallet pin from which an escape tooth drops as it leaves the pallet.

drop—The free, unrestrained motion of the escape wheel as it leaves one pallet pin before it drops upon the locking surface of another.

endshake—The free up and down space of pivoted wheels or arbors in their bearings.

engine turning—The decoration of a case or clock plate by means of circular curves.

escapement—The device with teeth or pivot pins by which the pendulum controls the rate of time keeping. It consists of an anchor and an escape wheel. An anchor is shaped so that it releases one tooth on a wheel at a given interval. This regulates the time and simultaneously gives an impulse to the pendulum.

escape wheel—A wheel at the end of the wheel train that is engaged by the anchor to regulate the clock's running.

faceplate—The faceplate in many instances is used primarily to hold the dial and keep it flat.

facing—The process of finishing the ends of pinion heads.

false plate—An intermediate plate between the movement and dial on some clocks to aid fitting the dial to the movement.

fan-fly—A fan with two blades used in clocks to

keep the interval between the strokes of the hammer uniform.

fly—A wind-resistant fan that regulates the speed of striking or chiming.

fork—A device suspended on the pendulum wire or a 400-day clock which transmits the impulse to the anchor pin from the pendulum.

four-hundred day clock—This is the type of clock, also known as an anniversary clock, that runs for a year on one winding and has a rotating pendulum on a long suspension spring, usually vibrating some eight times in a minute.

fourth wheel—Usually the wheel upon which is mounted the second hand.

grandfather clock—Strictly, a long-case clock.

great wheel—The first wheel in the going or striking train, nearest to the source of power, usually attached to the winding arbor and drum.

hairspring—The spiraled spring attached to the balance to govern the speed of the balance oscillations.

hands—The revolving indicators that point out the time.

horology—The science and study of time measurement.

hour wheel—A flat brass, toothed wheel mounted on a tube that fits over the cannon pinion and supports the hour hand.

impulse—The energy that must be delivered to an oscillator to keep it vibrating. It is one of the functions of the escapement to give impulse to pendulum or balance wheel. In many electric clocks, impulse is given by an electro-magnetically operated mechanism.

india stone—A fast-cutting artificial abrasive.

key, bench—A tool with varied size prongs capable of fitting into all sizes of winding arbors.

lantern pinion—A pinion consisting of two plates of brass or wood connected by cylindrical wires or trundles. Used in less expensive clocks.

leaf—The teeth of pinions are called leaves.

lever—Usually referred to as the pallet.

light-emitting-diode—A digital time display made up of lighted bars or dots.

liquid crystals—A digital display made up of bars or segments that reflect light with an electrical charge. Some liquid crystal displays allow light to pass through these segments when an electrical charge is activated.

mainspring—A long ribbon of steel used to supply the power for driving a clock.

mesh—The engagement or interlocking of a set of gear-teeth of one wheel with those of another.

micron—One-thousandth of a millimeter (0.001 mm). Generally used to describe the thickness of electroplating; most often as 20 microns gold. A millimeter measures 0.03937 inches.

minute wheel—The wheel in the dial train that connects the cannon pinion with the hour wheel.

moon dial—A dial, often found in the arch portion of a clock dial, that indicates the cycle of the moon.

motion train—A series of wheels that regulates the rotation of the hour and minute hands.

motion wheels—The gear wheels and pinions, generally between the dial and the front plate but often between plates, by which the revolutions of the central arbor are reduced to one revolution in 12 hours for the hour hand, with all hands turning in the same direction. Motion wheels are outside the main train that runs from mainspring or weight barrel through to the escapement.

motion work—The wheels used for causing the hour hand to travel 12 times slower than the minute hand.

movement—The term applied to the works of a clock as distinct from the case.

movement rest—A platform or vise upon which the movement is placed while it is being repaired.

oilstone—Generally, the Arkansas white stone used with oil.

oscillation—A cycle or double vibration; a swing of the pendulum or balance to both sides and back to the starting point.

pallets—The two projections from the ends of the anchor that engage with the escape-wheel teeth and allow one tooth to pass with each complete swing of the pendulum.

pawl—Another name for a dog, click, or ratchet. A projection, on a vertically moving part, that engages with a rotating part; or a rotating projection engaging with a vertically moving part. Principally escapement pallets that engage with scapewheel teeth, and gathering pallets that gather up striking and chiming rack teeth one by one.

pendulum—A body suspended from a fixed point called the escapement. It is attached by means of the crutch that controls the rate of time keeping.

pendulum spring—The ribbon of steel used to suspend the pendulum of a clock.

pillar—Posts of brass, steel, aluminum, or wood for keeping the plates of a clock in a fixed position as regards to each other. It connects to the front and back plates and establishes a fixed distance between them.

pinion—A small gear wheel, of not more than 20 teeth (called leaves), and most often of from 6 to 12 teeth. Pinions will be in one piece (of metal or synthetic material) or steel rods in brass endpieces ("lantern pinions"). Generally, they are solid and more often than not are part of their arbors rather than driven onto them. A pinion is usually driven by (meshes with) a larger wheel, except in the motion work and some electric movements.

pin pallet—The lever escapement wherein the pallet has upright pins instead of horizontally set jewels. Used in alarm clocks.

pivot—The finely turned end of an arbor that runs in a hole in the movement plate or a special cock.

plate—A sheet of metal, plastic, or even wood, two of which, with pillars at the four corners, form the housing for a clock's "train"—wheels, pinions, and arbors—that is pivoted between them.

rack and snail—An indexing system for striking that sets itself for correct striking shortly before striking begins.

ratchet—A saw-toothed wheel usually placed over a mainspring arbor and working with a retaining clock or pawl.

receiving pallet—The entrance pallet pin.

recoil click—A click designed so that it will not permit the mainspring to be wound dead tight, recoiling a bit after any winding.

rotor—The revolving part of an electric motor, particularly the unwound armature of a synchronous motor.

scapewheel—The wheel, farthest from the source of power, on which the pallets of an escapement act. The shape of the teeth varies according to the type of the escapement, but it has roughly the profile of a ratchet tooth.

screwplate—A steel plate with holes of many sizes threaded with cutting edges for the forming of small screws.

shake—The term used for freedom of a moving part in relation to a fixed part, most often of pivots in their holes.

snail—A cam, in shape like the profile of a snail's shell, used to regulate the number of blows struck at each hour. Often the 12 steps corresponding to the hours and on which the rack falls are cut on the snail.

spandrels—The decorated corners outside the dial.

staff—Another name for arbor or axle.

striking—Sounding of the hour, and possibly the half hour and quarters, from the same train on a bell or gong.

stud—A short metal rod of which the top is threaded to be screwed into a plate or similar piece.

third wheel—The train wheel between the center and fourth wheel.

train—A series of wheels and pinions through which power is transmitted from its source (usually weights or springs) to the escapement.

transistor—An electronic device capable of oscillating, or of acting as a relay or switch, when part of a suitable electrical circuit. Transistors can operate on a very low current, generate negligible heat, and they are very small. They are so indispensable in modern electronic circuits.

tripoli—A decomposed siliceous limestone used as a polishing powder; also known as rottenstone.

verge—An outmoded recoil frictional escapement with a crown escape wheel and pallets set at right angles to the axis of the escape wheel.

vibration—The swing of a pendulum or balance from one side to the other; half an oscillation.

wheel—A circular piece of metal on the perimeter of which are cut teeth.

works—The movement of a clock as distinguished from the case.

Bibliography

Bailey, Chris. *Two Hundred Years of American Clocks and Watches*. Englewood Cliffs, New Jersey: Prentice-Hall, 1975.

Bruton, Eric. *Antique Clocks and Clock Collecting*. London, England: The Hamlyn Publishing Group Limited, 1974.

———. *Dictionary of Clocks and Watches*. New York: Crown Publishers, 1963.

de Carle, Donald. *Complicated Watches and Their Repair*. New York: Crown Publishers, 1979.

———. *Watch and Clock Encyclopedia*. New York: Crown Publishers, 1977.

Fried, H.B. *Bench Practices for Watch and Clockmakers*. New York: Columbia Communications, 1974.

Gerschler, Malcolm C. *How to Train a Cuckoo (Clock)*. National City, California 92050: The Wag on the Wall, 1979.

Harris, H.G. *Advanced Watch and Clock Repair*. Buchanan, New York: Emerson Books, 1978.

———. *Handbook of Watch and Clock Repairs*. Buchanan, New York: Emerson Books, 1975.

Hope-Jones, F. *Electric Clocks and How to Make Them*. Watford, Herts, England: Argus Books, 1977.

Jagger, Cedric. *Clocks*. New York: Crown Publishers, 1973.

Kelly, Harold C. *Clock Repairing as a Hobby*. Chicago: Follett Publishing, 1972.

———. *Improving Your Clock Repairing Skills*. New York: Association Press, 1976.

Lloyd, H. Alan. *The Collector's Dictionary of Clocks*. Cranbury, New Jersey: A.S. Barnes and Company, 1964.

Marshall, Percival (ed. by). *Electric Clocks and Chimes*. Lings Langley, Herts, England: Argus Books, 1976.

Milham, Willis I. *Time and Timekeepers*. New York: The Macmillan Company, 1941.

Nicholls, Andrew. *Clocks in Color*. New York: Macmillan Publishing Company, 1975.

Pearson, Michael. *The Beauty of Clocks*. New York: Crown Publishers, 1978.

Schorsch, Anita. *American Clocks*. New York: Warner Books, 1981.

Schwartz, Marvin D. *Collectors' Guide, Antique American Clocks*. Garden City, New York: Doubleday & Company, 1975.

Shouffelberge, H. *Wheels and Pinions and How to Determine Their Exact Size*. Chicago: George K. Hazlitt and Company, 1977.

Smith, Eric. *Clock Repair and Maintenance*. New York: Arco Publishing Company, 1977.

———. *How to Repair Clocks*. Blue Ridge Summit, Pennsylvania 17214: TAB BOOKS, 1979.

Way, R. Barnard. *How To Make An Electric Clock*. Kings Langley, Herts, England: Argus Books, 1976.

Wels, Byron G. *How To Build Clocks and Watches*. Princeton, New Jersey: Auerbach Publishers, 1971.

Wood, Jr., Stacy B.C. and Stephen Kramer III *Clockmakers of Lancaster County and Their Clocks 1750-1850*. New York: Van Nostrand Reinhold Company, 1977.

Magazines

Bulletin of the National Association of Watch and Clock Collectors, Inc.
NAWCC Building
514 Poplar St.
Columbia, PA 17512

Watch & Clock Review
Golden Bell Press
2403 Champa St.
Denver, CO 80205

Index

Index

Edited by Steven Bolt